AF687728

Environmental Protection and Management

Environmental Protection and Management

Vivian Baldwin

MURPHY & MOORE
www.murphy-moorepublishing.com

Published by Murphy & Moore Publishing,
1 Rockefeller Plaza,
New York City, NY 10020, USA
www.murphy-moorepublishing.com

Environmental Protection and Management
Vivian Baldwin

© 2022 Murphy & Moore Publishing

International Standard Book Number: 978-1-63987-196-4 (Hardback)

This book contains information obtained from authentic and highly regarded sources. All chapters are published with permission under the Creative Commons Attribution Share Alike License or equivalent. A wide variety of references are listed. Permissions and sources are indicated; for detailed attributions, please refer to the permissions page. Reasonable efforts have been made to publish reliable data and information, but the authors, editors and publisher cannot assume any responsibility for the validity of all materials or the consequences of their use.

Trademark Notice: Registered trademark of products or corporate names are used only for explanation and identification without intent to infringe.

Cataloging-in-Publication Data

Environmental protection and management / Vivian Baldwin.
p. cm.
Includes bibliographical references and index.
ISBN 978-1-63987-196-4
1. Environmental protection. 2. Environmental management. 3. Environmental quality.
4. Environmental engineering. 5. Environmental sciences. I. Baldwin, Vivian.
TD170 .E58 2022
363.7--dc23

TABLE OF CONTENTS

PREFACE

The purpose of this book is to help students understand the fundamental concepts of this discipline. It is designed to motivate students to learn and prosper. I am grateful for the support of my colleagues. I would also like to acknowledge the encouragement of my family.

The environment is degraded due to various reasons such as population growth, pollution, overconsumption, etc. The practices aimed at conserving and reducing these risks within the environment fall under environmental protection. Environmental management is a dynamic concept which is concerned with the management of environment using different tools and techniques such as environmental accounting, policy, auditing etc. Some of the goals of environmental protection are ensuring sustainable use of resources, conserving biological diversity, controlling land degradation, etc. The primary approaches to environmental protection and management are voluntary and international environmental agreements, and ecosystems approach. This book elucidates new techniques and their applications in a multidisciplinary approach. The topics included herein are of utmost significance and bound to provide incredible insights to readers. For all those who are interested in environmental protection and management, this textbook can prove to be an essential guide.

A foreword for all the chapters is provided below:

Chapter – Introduction

Environmental protection and management deals with the conservation of natural resources and the environment by adopting certain practices. It includes the use of environmental engineering and technology, sustainable land management, watershed management, environmental monitoring, etc. This chapter has been carefully written to provide an easy understanding of environmental protection and management.

Chapter – Key Concepts in Environmental Protection and Management

Environmental protection and management incorporates a number of concepts such as sustainable and environmental management, ecosystem services, habitat conservation, life-cycle assessment, waste minimisation and environmental management system, etc. This chapter closely examines these key concepts of environmental protection and management to provide an extensive understanding of the subject.

Chapter – Natural Resource Management

The management of natural resources like soil, water, land, plants and animals is referred to as natural resource management. It includes wildlife management, fisheries management, natural resource economics, integrated landscape management, etc. All these methods related to natural resource management have been thoroughly discussed in this chapter.

Chapter - Water and Soil Conservation

Water conservation refers to the activities and strategies towards the sustainable management of natural resources of water. The utilization of methods to prevent soil loss from erosion, acidification and salinization is known as soil conservation. The topics elaborated in this chapter will help in gaining a better perspective about the diverse aspects of water and soil conservation.

Chapter - Environmental Pollution Management

Environmental pollution management refers to the control of emissions and effluents into soil, air and water. It includes the areas of air pollution control, sewage treatment, industrial wastewater treatment and noise pollution control. This chapter closely examines these key concepts of environmental pollution management to provide an extensive understanding of the subject.

Vivian Baldwin

Introduction

1

- **Environmental Protection**
- **Environmental Management**
- **Conservation of Biodiversity**
- **Sustainable Land Management**
- **Watershed Management**
- **Environmental Monitoring**
- **Globalization and its Impact on the Environment**
- **Environmental Engineering**
- **Environmental Technology**

Environmental protection and management deals with the conservation of natural resources and the environment by adopting certain practices. It includes the use of environmental engineering and technology, sustainable land management, watershed management, environmental monitoring, etc. This chapter has been carefully written to provide an easy understanding of environmental protection and management.

Environmental Protection

Environmental protection refers to any activity to maintain or restore the quality of environmental media through preventing the emission of pollutants or reducing the presence of polluting substances in environmental media. It may consist of:

- Changes in characteristics of goods and services.

- Changes in consumption patterns.
- Changes in production techniques.
- Treatment or disposal of residuals in separate environmental protection facilities.
- Recycling.
- Prevention of degradation of the landscape and ecosystems.

Importance of Environmental Protection

In recent decades, many environmental problems have increased as the result of human activities and unplanned management of the technological development those interference ecosystems. Therefore, a dispute between the importance of conservation and preservation of ecosystems to protect environment and the necessity to satisfy human desire by sacrifice the environment has been arise across the world. According to Glossary of Environment Statistics the term "environmental protection" can be defined as the prevention to conserve and preserve the standard healthy level of environmental media by reducing the production of pollutants or polluting substances in environmental media.Various human activities have induce many undesirable effects to the environment which can be threatening human health, economic, natural resources and gene pool of ecosystems such as pollutions, greenhouse effect, global warming and soil erosion. Firstly, the environmental pollution is one of the main reasons why we should fight to protect environment. Besides, global warming is also another reason caused by the deforestation. Furthermore, warm climate change and flood also increase the opportunity of spread out pests and vector diseases.

Pollution can be divided into four types; water pollution, air pollution, thermal pollution and sound pollution. Fossils fuels used in most factories, petroleum and gas usage for vehicle is the major cause of air pollution. Environment pollution can also be an addition harmful substance that could affect human health and human life. Given the above, environmentalist should fight for a better environment because pollutions caused by burning of fossil fuels will results in severe environmental problems such as the occurrence of acid rain due to the production of sulphur dioxide (SO_2) and nitrogen monoxide (NO). Casiday and Frey mentioned that combustion process not only increases the concentration of carbon dioxide (CO_2) in the atmosphere but it is also the main source in producing high level of nitrogen monoxide (NO) and sulphur dioxide (SO_2). These are the two major reactants that cause the formation of acid rain. Acidity of acid rain changes the pH of the river water and lakes which then disrupt the nature habitat of aquatic organisms and reduce the chance of survival of aquatic organisms, for example fishes cannot survive due to lack of oxygen. In addition, the insoluble aluminum ions added to the water can causes water to become poisonous which can also be known as water pollution. For example, Norway has blamed that the sea and lakes are poisoned by the formation of acid rain for many years due to the unclean air pollution that comes from Britian's power stations.

Since the global concentration of carbon dioxides have increased given the reasons of combustion of fossil fuels and also human activities such as deforestation. The reasons for deforestation are usually resulted from cutting down the forests for lumber logging and also for building a new farming for animal. Forest can also be known as "carbon sink" because trees can be used as an absorber of carbon dioxide in the environment and in returns release some oxygen to the atmosphere. However, unplanned deforestation activities have significantly reduced the concentration of oxygen (O_2) and caused the rise in temperature of the Earth. Christopher Monckton also emphasises that every doubling the concentration of carbon dioxide in the atmosphere is sufficient to rise the global surface temperature by 3.26 °C. Rise in global temperature will significantly caused ice melting, for instance, according to the My Eco Project Organization, the reports of ice melting at Arctic Retreat prepares in the late summer 2007 surprised the experts that every week, the Arctic sea ice is continuously melting and the amount of melted ice can fill up to as large as two Britain country. Meanwhile, the consequence of ice melting at Arctic has increased the sea water levels. According to the evidence gathered by Shepherd and his team, they found that the sea water level has increased by 2.6% which has the same volume with 49 microns per year spread across the oceans from over the world due to the distinct value between the density and temperature of ice and sea water. In addition, the polar bear will also face extinction in the future due to global warming. Based on the report from National Geographic News, several studies that have been conducted by the U. S. government has show that melting of Arctic's ice that caused by the global warming will endangered two-thirds of the world's polar bears and cause them to face extinction by 2050. Given the above fact, environmentalist should fight for a better environment and save our earth.

Furthermore, environmental protection should be implemented due to warm climate changes and flood which heighten the risks of spread out pests and vector diseases. Some infectious diseases such as malaria, dengue, cholera and encephalitis can also spread out rapidly throughout the whole area by mosquitoes, flies and other insects those usually adapt to live in warm weather region. As World Health Organization notes that contamination of water resources due to the occurrence of flood enlarge the chances of getting water-borne diseases and also for mosquitoes to carry disease around the environment. Malaria and dengue fever have threaten the Southeast Asia and South Pacific island due to the climate change as increases the population of mosquitoes and migration of refugees. Hence, it can be concluded that flood and climate change will destroy our safety health environment. Moreover, high temperature in the river will encourage the growth of algal broom and causes the water pollution index to increases and decrease the amount of oxygen supplied for the aquatic organisms. Since the river water pollution increases dramatically, the level of biochemical oxygen demand which known as BOD level also will increases. The higher the biochemical oxygen demand level will result in the higher of the pollution level. For instance, large amount of blue-green algae devastate the nature of the universal solvent and threaten the public health by releasing toxins into the water. Diseases and infection that can cause by the algae's toxin are sore throats, gastro-enteritis and skin or eye-infections.

Environmental Management

The ecological balance and ecosystem stability are duly maintained by the nature itself but the emergence of modern industrial era has disturbed the ecological balance through heavy industrialization, technological revolution, faster growth of means of transportation, rapacious exploitation of resources, unplanned urbanization etc.

In other words, the anthropogenic activities of modern 'economic and technological' man have disturbed the harmonious relationships between the environment and human beings. Environmental management is thus, the process to improve the relationship between the human beings and environment which may be achieved through check on destructive activities of man, conservation, protection, regulation and regeneration of nature.

The process, environmental management is related to the rational adjustment of man with nature involving judicious exploitation and utilization of natural resources without disturbing the ecosystem balance and ecosystem equilibrium.

If the natural resources are overexploited, it will affect socio-economic development of a nation. Thus, environmental management must take into consideration the ecological principles and socioeconomic needs of the society i.e., it involves socio economic developments on one hand and maintenance of environmental quality on other hand.

From the above discussion, it is clear that environmental managements has two major aspects:

- Socio-economic development.
- Stability of biosphere in general and stability of individual ecosystems in particular.

Scope and aspects of Environmental Management

Environmental management is very wide in scope and includes all the technical, economical and other aspects of environment.

The broader objectives of environmental management includes:

- To identify the environmental problem and to find its solution.
- To restrict and regulate the exploitation and utilization of natural resources.
- To regenerate degraded environment and to renew natural resources (renewable).
- To control environmental pollution and gradation.
- To reduce the impacts of extreme events and natural disaster.

- To make optimum utilization of natural resources.
- To assess the impacts of proposed projects and activities on environment.
- To review and revise the existing technologies and make them ecofriendly.
- To formulate laws for the implementation of environmental protection and conservation programmes.

The components of environmental management are based on five fundamental aspects:

Environmental Perception and Public Awareness

The environmental perception and public awareness considers the following points:

- Sources of environmental perception and public awareness.
- Level of environmental perception.
- Role of environmental perception in environmental planning and management.

Environmental Education and Training

Environmental education and training should be given at school, college and University levels by professionals.

Resource Management

The resource management considers the following points:

- Classification of natural resources.
- Survey and evaluation of ecological resources.
- Preservation of resources.
- Conservation of resources.

Control of Environmental Degradation and Pollution

The environmental degradation and pollution can be checked by considering the following points:

- Control of environmental degradation and pollution.
- Adopting suitable preventive mechanisms to reduce natural hazards and disaster.
- Regeneration of degraded environment.

Environmental Impact Assessment

The environmental impact assessment involves:

- Appraisal of existing environmental conditions.
- Appraisal of existing and proposed production methods.
- Methologies and procedures.
- Probable impacts of existing and proposed project.
- Review of technology and required improvement.

Characteristics and Objectives

Environmental management is the process of allocating natural and man-made resources so as to make optimum use of the environment in satisfying not only the present basic human needs but of the coming generations also.

This management implies an element of conscious choice from a variety of alternative proposals and furthermore that such a choice involves purposeful commitment to recognised and desired objectives.

Environmental management is not merely a management of environment but it is essentially the management of various activities with intolerable constraints imposed by the environment itself and with full consideration of ecological factors. Thus, it involves environmental planning, conservation of resources, environmental status evaluation, and environmental legislation and administration.

The focus of environmental management is on implementation, monitoring and auditing; on practice and coping with real-world issues, rather than theoretical planning. A close integration with environmental planning is desirable.

Thus, as stated earlier, environmental management is a field of study dedicated to understanding human-environment interactions and the application of science and common-sense to solving problems.

The characteristic features of the environmental management are:

- It deals with a world affected by humans.
- It supports sustainable development.
- It demands a multidisciplinary approach.
- It has to integrate different development viewpoints.
- It concerns with short-term and long-term planning as well as from local to global scale.
- It seeks to integrate natural and social science, policy making and planning.

During the last three decades too much awareness has been developed regarding environmental protection and quality of life. The dictionary of environment is renewed regularly with new terminologies like clean technology, environmental auditing, environment-friendly products, environmental impact assessment, environmental resource conservation, etc.

But all these aspects have been converged when the wider concept of environmental management has been emerged and also accepted as a tool for sustainable development. Environmental management, as defined in Goudie, "provides resources from the bioenvironmental systems of the planet but simultaneously tries to retain sanative, life-supporting ecosystems. It is therefore an attempt to harmonise and balance the various enterprises for his own benefit."

Time has now come when our policy makers as well as society should aim to protect, conserve and regulate the development in such a way that it will not create any adverse effect on ecosystem and needs of the people can also be fulfilled. Throughout the world, particularly in developing countries, these are an urgent need for the management of the total environment. In the first instance environ- mental management must do three things:

- Identify goals.
- Establish whether these can be met.
- Develop and implement means to do what it deems possible. A simple scheme for environmental management has been depicted in figure.

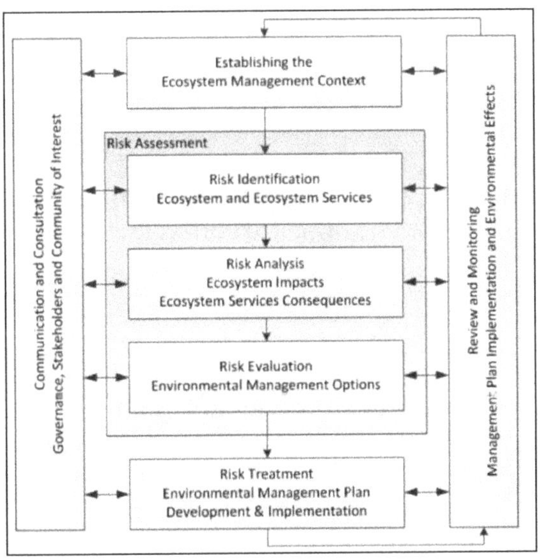

A General Scheme for Environmental Management.

Thus, environmental management is an approach which integrates ecology, policy making, planning and social development.

Its main objectives include:

- To prevent and solve environmental problems.
- To establish limits.
- To develop research institutions and monitoring systems.
- To warn threats and identify opportunities.
- To suggest measures for resource conservation.
- To develop a strategy for the improvement of quality of life.
- To suggest long-term and short-term policies for sustainable development.
- To identify new technology for sustainable development.

In brief, environmental management is necessary for environmental planning which implies the optimal utilisation of the earth's resources and preservation of the quality of environment for the healthy growth of society.

The term 'sustainable development' was first used at the time of the Cocoyoc Declaration on Environment and Development in early 1970. Since then, it has become a matter of great concern not only to the international and national organisations but also to the policy makers.

In 1987, the World Commission on Environment and Development has published a report Our Common Future emphasising the need for sustainable development and defined it as: "Sustainable development is the development that meets the needs of the present without compromising the ability of future generations to meet their own needs". Others refer to it as "improving the quality of human life while living within the carrying of the supporting ecosystem".

Now, sustainable development, also referred as 'eco-development', has been accepted as a concept of development by all the nations of the world. Agenda 21, adopted in the UNCED at Rio de Janeiro, is a programmer of action for sustainable development, also known as 'Rio Declaration' on Environment and Development.

Approaches to Environmental Management

The best approach to environmental management is an integrated approach in which all the components of the environment are taken into consideration and its proper management, as a whole, is done.

While doing so, the following aspects should be taken into account:

Perception and Awareness of Environment

- Source of perception and awareness.

- Level of perception.
- Role of perception in environmental planning.

Environmental Education and Training

- At school, college and university levels.
- Through media.
- At research and training institutes.

Resource Management

- Classification of natural resources.
- Survey and evaluation of eco-resources.
- Conservation of energy, mineral, forest, soil and water resources.
- Proper utilisation of human resources.

Environmental Impact Assessment

- Assessment of present conditions.
- Assessment of the impact of industrial and technological developments.
- Assessment of the efforts done for eco-balance.
- Development of eco-friendly techniques.

Control Over Environmental Degradation and Pollution

- Purification of degraded environment.
- Pollution control.
- Monitoring.
- Forecasting of natural hazards and to minimise the losses.

There are two approaches of managing the environment. (i) Preservative approach, and (ii) conservative approach. According to the first approach, man should not disturb the natural system and should adapt to it. But this is not possible because for all types of developmental activities he will have to use nature and its components, resulting in eco-imbalance of varied intensity.

The conservative approach is that there should not be overexploitation of nature and conservation of natural resources is essential for sustainable development.

In fact, proper utilisation and conservation of resources is the prime objective of environmental management. In the World Conservation Strategy, the three main objectives stated were: the maintenance of essential ecological processes, the preservation of genetic diversity, and the sustainable utilisation of natural resources.

In recent years, several approaches have been developed for the management of environment. Regional planners often adopt a human ecology approach, while other planners suggest a system analysis or an ecosystem approach. Doxiadis has developed a science of planning settlement in balance with nature and termed it as ekistics.

The various approaches developed for the environmental management are:

- Ad hoc approach: Developed in reaction to a specific situation.
- Problem-solving approach: For identification of problems and needs and implement solutions.
- Systems approach: Such as ecosystem, agro-ecosystem, etc. Regional approach: based on ecological zones such as watershed, river basin, coastal zone, command area development, island, etc.
- Specialist discipline approach, often adopted by professionals for air, water and land management, urban management, tourism management, and environmental health.
- Voluntary sector approach encouraged and supported by NGOs.
- Commercial approach: For environmental management for business.
- Human ecology approach: For study of relations between humans or society and nature.
- Political ecology approach: To develop policies and laws.

Apart from the above, an environmental management system (EMS) approach has been developed as an integrated and proactive approach to environmental issues.

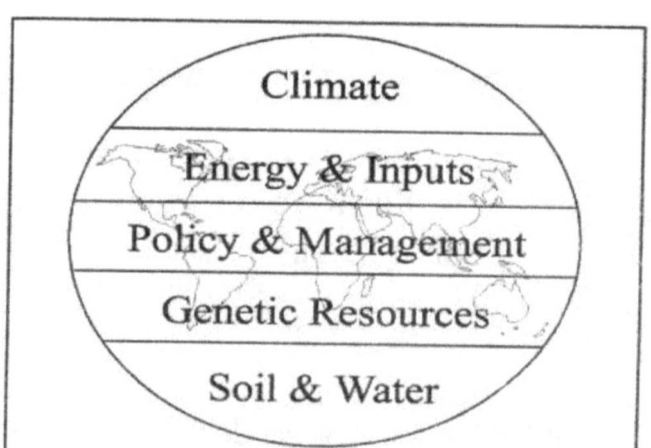

Environmental Management System.

It helps industry or other bodies and is designed to ensure that an environmental policy and environmental objectives are adopted and followed. Illustrates a basic EMS approach developed by Hunt and Johnson.

Thus, the EMS system approach helps to:

- Develop a proactive environment approach.
- Ensure a balanced view across all functions.
- Enable effective, directed environmental goal setting.
- Make the environmental auditing process effective.

Ecosystem Approach

The ecosystem concept has become a widely used conceptual tool for research after 1945. Nowadays, ecologists often adopt an ecosystem approach when seeking to understand and monitor a given situation. The ecosystem concept allows the environmental managers to look at portions of complex nature as an integrated system. It may be applied to cities or agriculture (urban ecosystems and agro-ecosystems respectively).

The ecosystem approach allows a holistic view of how components work together, in other words, it can incorporate human dimensions into biosphere functioning. It also helps define the temporal and spatial scale of management, and thus, is a multidisciplinary approach in order to deal with complexities of ecosystem function and usage.

Human Ecology Approach

Human ecology is the study of relations between humans or society and nature, through a multidisciplinary approach. Similar to that of ecosystem approach, the main emphasis in this approach is on social relationship with environment which is a primary aspect of any management for planning and development. The scale of approach may be local to global, and it supports holistic study.

The concept of 'Social Impact Assessment' (SIA) seeks to assess whether a proposed development alters quality of life and sense of well-being and how individuals, groups and communities will adopt to change caused by development.

The socio-economic and biophysical aspects of the environment are interconnected, therefore, for environmental management; human ecology approach should be adopted along with political ecology approach.

Political Ecology Approach

Political ecology also studies relationships between society and nature. It holds that radical changes in human habits are required in order to counter environmental degradation and achieve sustainable development.

These are likely to be different perceptions of environmental needs and problems between planners, policy makers, ministers, various departments of government, etc. All this can be effectively done by adopting political ecological approach of environment management.

Commercial Approach

Nowadays, there has been an increasing emphasis on environmental management for business as well as role of business houses in environmental protection.

The emphasis is on:

- Green corporate environmental management.
- Green business ethics.
- Eco-auditing.
- Impact assessment, hazard and risk assessment.
- Green marketing, labelling.
- Recycling and waste disposal.
- Rnvironmentally sound investment and funding.
- Total quality management.

The commercial approach involves corporate priority, employee education, customer advice, transfer of technology, prior assessment, facilities and operations, research and compliance and reporting.

Conservation of Biodiversity

Conservation of biological diversity is essential for the survival of the human race.

Objectives and Advantages of Biodiversity Conservation

- Conservation of biological diversity leads to conservation of essential ecological diversity to preserve the continuity of food chains.
- The genetic diversity of plants and animals is preserved.
- It ensures the sustainable utilisation of life support systems on earth.
- It provides a vast knowledge of potential use to the scientific community.

- A reservoir of wild animals and plants is preserved, thus enabling them to be introduced, if need be, in the surrounding areas.
- Biological diversity provides immediate benefits to the society such as recreation and tourism.
- Biodiversity conservation serves as an insurance policy for the future.

Types of Conservation

Ex Situ Conservation

Conserving biodiversity outside the areas where they naturally occur is known as ex situ conservation. Here, animals and plants are reared or cultivated in areas like zoological or botanical parks.

Reintroduction of an animal or plant into the habitat from where it has become extinct is another form of ex situ conservation. For example, the Gangetic gharial has been reintroduced in the rivers of Uttar Pradesh, Madhya Pradesh and Rajasthan where it had become extinct.

Seedbanks, botanical, horticultural and recreational gardens are important centres for ex situ conservation.

In Situ Conservation

Conserving the animals and plants in their natural habitats is known as in situ conservation. This includes the establishment of:

- National parks and sanctuaries.
- Biosphere reserves.
- Nature reserves.
- Reserved and protected forests.
- Preservation plots.
- Reserved forests.

Agrobiodiversity Conservation

After the introduction of cotton, tobacco, sugarcane, sunflower, soyabean and so on, farmers became victims of monocultures in their greed for money. Therefore many of the indigenous varieties of crops were lost. Moreover, the hybrid varieties of fruits and vegetables (e.g. tomatoes), introduced for pulp are more susceptible to disease and pests. Though hybrid varieties are preferred, traditional wild varieties of the seeds should be conserved for future use in the event of an epidemic which would completely wipe out the hybrids.

Botanical gardens, agricultural departments, seed banks etc., alone should not be given the responsibility of agrobiodiversity conservation. Every farmer, gardener an cultivator should be aware of his role in preserving and conserving agrobiodiversity.

Convention of Biological Diversity

The aim of the convention is to save species and plants from extinction and their habitats from destruction.

The developed countries are looking for a sustainable supply of biological resources from the developing countries and easy access to them as well. The developing countries lacking the technology to exploit their resources are inviting the developed countries to do so. This has resulted in the developed nations channeling out the benefits of these natural resources. The developing countries are now demanding a higher share of the accrued economic benefits. The developed nations are also concerned by the unsustainable exploitation of natural wealth, particularly rainforests.

Key Points from the Convention on Biological Diversity

The aim of the Convention on Biological Diversity is 'the conservation of biological diversity, the sustainable use of its components and the fair and equitable sharing of the benefits arising out of the utilization of genetic resources. The convention stipulates that Parties must:

- Develop national strategies for the conservation and sustainable use of biological resources.
- Establish protected areas, restore degraded ecosystems, control alien species, and establish ex-situ conservation facilities.
- Establish training and research programmes for the conservation and sustainable use of biodiversity and support such programmes in developing countries.
- Promote public education and awareness of the conservation and sustainable use of biodiversity.
- Recognize the right of governments to regulate access to their own genetic resources, and, wherever possible, grant other Parties access to genetic resources for environmentally sound uses.
- Encourage technology and biotechnology transfer particularly to developing countries.
- Establish an information exchange between the parties on all subjects relevant to biodiversity.

- Promote technical and scientific cooperation between parties (particularly to developing countries) to enable them to implement the convention.
- Ensure that countries that provide genetic resources have access to the benefits arising from them.
- Provide financial resources to developing countries/parties to enable them to carry out the requirements of the convention.

Major Causes for Biodiversity Loss

Loss of biodiversity occurs when either the habitat essential for the survival of a species is destroyed, or particular species are destroyed. The former is more common as habitat destruction is a fallout of development. The latter reason is encountered when particular species are exploited for economical gain or hunted for sport or food.

Extinction of species may also be due to environmental factors like ecological substitutions, biological factors and pathological causes which can be caused by nature or man.

Natural Causes for the Loss of Biodiversity

Natural causes include floods, earthquakes, landslides, natural competition between species, lack of pollination and diseases.

Man-made Causes for the Loss of Biodiversity

- Destruction of habitat in the wake of developmental activities like housing, agriculture, construction of dams, reservoirs, roads, railway tracks, etc.
- Pollution, a gift of the industrial revolution can be given the pride of place for driving a variety of species in air, water and land towards extinction.
- Motorcars, air-conditioners and refrigerators, the three symbols of a modern, affluent society, have been instrumental in global warming and ozone depletion. They have drastically altered the climate with disastrous effects on the various species. Factories and power stations spewing out poisonous gases and effluents have fouled up the environment bringing death and disease to many species. Oil spills and discharge of sewage have ravaged the oceans and coastal habitats.
- A large number of species are threatened by overhunting, poaching and illegal trade.
- Indiscriminate use of toxic chemicals and pesticides and overexploitation of wildlife resources for commercial purposes are responsible for the rapid decline in the number of some species. The tiger for instance is hunted for its claws and

other parts believed to be effective cures for various ailments of man. Snakes and crocodiles are killed in large numbers for their skin and minks, sable, ermine, etc., are in demand for the luxury and warmth of their fur.

- Genetic erosion arises from the loss (due to commercial and anthropogenic pressures) of habitats rich in biodiversity and from the disappearance of the traditional conservation practices of wild species in their habitats by rural and tribal people.

Sustainable Land Management

Land provides an environment for agricultural production, but it also is an essential condition for improved environmental management, including source/sink functions for greenhouse gasses, recycling of nutrients, amelioration and filtering of pollutants, and transmission and purification of water as part of the hydrologic cycle. The objective of sustainable land management (SLM)1 is to harmonise the complimentary goals of providing environmental, economic, and social opportunities for the benefit of present and future generations, while maintaining and enhancing the quality of the land (soil, water and air) resource. Sustainable land management is the use of land to meet changing human needs (agriculture, forestry, conservation), while ensuring long-term socioeconomic and ecological functions of the land.

Sustainable land management is a necessary building block for sustainable agricultural development, and it is a key element in AGENDA 21's goal of sustainable development. Sustainable agricultural development, conservation of natural resources, and promoting sustainable land management are key objectives of the new World Bank rural investment program, From Vision to Action, and increasingly these objectives are being included in all agricultural development and natural resources management projects.

Sustainable land management combines technologies, policies, and activities aimed at integrating socioeconomic principles with environmental concerns, so as to simultaneously:

- Maintain and enhance production (productivity).
- Reduce the level of production risk, and enhance soil capacity to buffer against degradation processes (stability/resilience).
- Protect the potential of natural resources and prevent degradation of soil and water quality (protection).
- Be economically viable (viability).
- Be socially acceptable, and assure access to the benefits from improved land management (acceptability/equity).

The definition and these criteria, called pillars of SLM, are the basic principles and the foundation on which sustainable land management is being developed. Any evaluation of the sustainability has to be based on these objectives: productivity, stability/resilience, protection, viability, and acceptability/equity. The definition and pillars have been field tested in several countries, and they were judged to provide useful guidance to assess sustainability.

The lack of a comprehensive, quantifiable definition for sustainable land management is sometimes considered to be a serious deficiency. Yet, as argued by Gallopin, a research model for sustainability has to be more flexible and therefore less easy to quantify than a research model for chemistry, physics, or classical agronomy. Such a research model must be designed around an evaluation process (rather than thematic context), because it is intended to test the likelihood of certain events taking place and the aggregate impacts of these events, rather than specifics of various null hypothesis or the impacts of certain inputs or land management interventions. Essentially the research model must include a goal statement, a conceptual framework, a set of procedures, and criteria (indicators) for diagnosis. One of the main objectives of such a research model is to evaluate the impacts of events which are uncertain, but the process of evaluation is guided by scientifically defined protocols.

Principles and Criteria for Sustainable Land Management

Experiences gained from field projects in developing and developed countries has identified a series of principles for sustainable land management, and these can be used as general guidelines for development projects. The most useful of these are summarised below:

Global Concerns for Sustainability

- Sustainability can be achieved only through the collective efforts of those immediately responsible for managing resources. This requires a policy environment that empowers farmers and other, local decision makers, to reap benefits for good land use decisions, but also to be held responsible for inappropriate land uses.

- Integration of economic and environmental interests in a comprehensive manner is necessary to achieve the objectives of sustainable land management. This requires that environmental concerns be given equal importance to economic performance in evaluating the impacts of development projects, and that reliable indicators of environmental performance be developed.

- There is urgent need to resolve the global challenge to produce more food to feed rapidly rising global populations, while at the same time preserving the biological production potential, resilience, and environmental maintenance systems of the land. Sustainable land management, if properly designed and implemented, will ensure that agriculture becomes a part of the environmental solution, rather than remaining an environmental problem.

Sustainable Agriculture

- More ecologically balanced land management can achieve both economic and environmental benefits, and this must be the foundation (linch pin) for further rural interventions (investments). Without good land management, other investments in the rural sector are likely to be disappointing2. At the same time, arguing for the continued maintenance of agriculture without reference to environmental sustainability is increasingly difficult. Indicators of land quality are needed to guide us along the way.
- Agricultural intensification is often necessary to achieve more sustainable systems. This requires shifts to higher value production, or higher yields with more inputs per unit of production and higher standards of management (more knowledge intensive). However, sustainable agriculture has to work within the bounds of nature not against them. Many yield improvements can be achieved by optimizing efficiency of external inputs rather than trying to maximize yields.
- The importance of off-farm income should not be underestimated because it i) supplements cash flow on the farm, ii) generates an investment environment for improved land management, and therefore iii) reduce production pressures on land.

Sharing Responsibilities for Sustainability

- Farmers and land managers must expand their knowledge of sustainable technologies and implement improved procedures of land stewardship. The preferred option is not to tell the farmer what to do (command and control legislation), but to create an enabling environment through policy interventions where farmers are more free to make the right choice. A policy environment where farmers are more empowered, but also held accountable, for achieving the objectives of sustainable land management is essential. However, sustainable land management is the responsibility of all segments of society. Governments must ensure that their policies and programs do not create negative environmental impacts, and society needs to define requirements for land maintenance and develop a "social" discount rate for future land use options that encourages the most sustainable use.
- Concerns for sustainable land management go beyond agriculture to include the legitimate interests of other aspects of land stewardship, including wildlife, waterfowl and biodiversity management. There is increasing evidence that society is demanding that farmers become stewards of rural landscapes, and that agriculture become more than simply putting food on the table. Many of society's environmental values may not represent economic gains for farmers, however, and farmers cannot shoulder all the costs of environmental maintenance.

Relationships among Soil Quality, Land Quality and Sustainable Land Management

New concepts of soil and land quality are emerging, and often these are used interchangeably. These concepts and their relationships are summarized below, to the extent that some concensus is available on how these should be applied:

Soil quality is the capacity of a specific soil to function within natural or managed ecosystem boundaries to sustain plant and animal production, maintain or enhance water quality, and support human health and habitation.

Land quality is the condition, state or "health" of the land relative to human requirements, including agricultural production, forestry, conservation, and environmental management.

Sustainable land management combines technologies, policies, and activities aimed at integrating socio-economic principles with environmental concerns so as to simultaneously maintain or enhance production, reduce the level of production risk, protect the potential of natural resources and prevent (buffer against) soil and water degradation, be economically viable, and be socially acceptable.

These concepts span the scales of detail, application, and levels of integration with socio-economic data. Soil quality is the most restrictive, followed by land quality and then sustainable land management. Soil quality is effectively a condition of a site, and it can be studied using soil data alone. Land quality requires integration of soil data with other biophysical information, such as climate, geology and land use. Land quality is a condition of the landscape, i.e. it is a biophysical property, but includes the impacts of human interventions (land use) on the landscape. Sustainable land management requires the integration of these biophysical conditions, i.e. land quality, with economic and social demands. It is an assessment of the impacts of human habitation, and a condition of sustainable development.

These are more than simple differences in semantics; the concepts differ in the kinds and scale of the processes being described, the data used for input, and the amount and kinds of integration with other disciplines. However, the concepts form a continuum over the landscape, and they must be applied for different types and scales of land use.

Land Quality Indicator Program

Assessment of sustainable land management requires appropriate evaluation instruments, such as Land Quality Indicators. However, land quality, like the concept of sustainable land management of which it is a part, requires operational definitions and specific, measureable indicators if it is to be more than an attractive, conceptual phrase.

The World Bank, in collaboration with UNEP, UNDP, FAO and the CGIAR, is developing a program called Land Quality Indicators (LQIs), as a means to better coordinate actions on land related issues such as land degradation. In the area of economic and social data, and in some cases for air and water quality, indicators are already in regular use to support decision-making at global, national and sub-national levels. In contrast, few such indicators are available to assess, monitor and evaluate changes in the quality of land resources. Land refers not just to soil but to the combined resources of terrain, water, soil and biotic resources that provide the basis for land use. Land quality refers to the condition or "health" of land, and specifically to its capacity for sustainable land use and environmental management.

The LQI program addresses the dual objectives of environmental monitoring as well as sector performance monitoring for managed ecosystems (agriculture, forestry, conservation and environmental management). It is being developed for application at national and regional scales, but it is also part of a larger, global effort on improved natural resources management. The LQI program is in response to the United Nations Conference on Environment and Development, and it fits with Agenda 21 expectations as well as the Convention to Combat Desertification.

Agricultural cropland, including agroforestry, as well as forested lands, range and pasture lands, are under increasing pressure because of population migrations to marginal land areas and agricultural intensification on existing cultivated lands. Sustainable land use intensification requires the maintenance or enhancement of the productive potential of the land resources, i.e. increases in food supplies must come from agricultural intensification rather than from area expansion, but this must be done without degrading the land resource on which production depends. The question is how this should be achieved, and how to monitor progress towards this objective in the different agroecological regions of developing countries. LQIs are tools to help us along the way.

In general and particularly in developing countries, it is essential that scarce resources devoted to land management be used more cost-efficiently and that policy-makers have at least rough indicators of whether environmental conditions and land quality are getting better or worse. Land quality indicators, such as nutrient balance, loss of organic matter, land use intensity and diversity, and land cover are useful to task managers and decision makers to monitor and improve the performance of projects with respect to their socio-economic and environmental impacts, and to assess the trend towards or away from land use sustainability. While routine project performance indicators based on cost-benefit analyses (input-output factors, risk and economic performance indicators), are necessary to monitor the activities and components of a project, LQIs are required to evaluate the environmental impact(s). The quantitative assessment of physical impacts, such as depletion of soil nutrients, loss of organic matter, soil erosion, water contamination etc. may appear to be costly and cumbersome during project implementation, but the long-term negative impact of reduced land quality, such as decreased efficiency of fertilizers, increased erosion, increased fuel consumption, increased pest infestation (nematodes, etc.), often result in rehabilitation costs that are much higher. The LQI approach focuses on preventive maintenance rather than rehabilitation, and provides the methodology and the approach to integrate the socio-economic and biophysical information that are required for better informed sustainable land management strategies.

The development of LQIs follows a logical framework, providing information not only on the state of the resources, but also the underlying causes (or "pressure") as well as the response of the society to the state and the pressure exerted on the land resources. Because of the nature and the complexity of land issues, the LQI program recommends addressing issues of land management by agro-ecological zones (Resource Management Domains or "Terroirs"). This new approach puts the focus on evaluating impacts of human interventions on specific landscapes, rather than emphasizing only biophysical variables as was the case in the past. At the same time, this spatial stratification favors incorporating farmer and other local knowledge into the overall process of improved agricultural and environmental land management.

A research strategy for the Land Quality Program was developed during a two day research planning meeting, Washington, D.C., 1996, sponsored by the World Bank. A panel of internationally acclaimed scientists and administrators established the objectives and priorities for the research, defined strategic alliances to be developed with ongoing national and international programs, and identified potential sources of funding. They also achieved international agreement on a core set of strategic land quality indicators.

Core LQIs for managed ecosystems (agriculture and forestry) in the major agro-ecological zones (AEZs) of tropical, sub-tropical and temperate environments, and recommended for development in the short term include:

- Nutrient balance: This describes nutrient stocks and flows as related to different land management systems used by farmers in specific AEZs and specific countries.
- Yield trends and yield gaps: This describes current yields, yield trends, and actual:potential, farm level yields for the major food crops in different countries.
- Land use intensity: Describes the impacts of agricultural intensification on land quality. Intensification may involve increased cropping, more value-added production, and increased amounts and frequency of inputs; emphasis is on the management practices adopted by farmers in the transition to intensification.
- Land use diversity (agro-diversity): Describes the degree of diversification of production systems over the landscape, including livestock and agro-forestry systems; it reflects the degree of flexibility (and resilience) of regional farming systems, and their capacity to absorb shocks and respond to opportunities.
- Land cover: Describes the extent, duration and timing of vegetative cover on the land during major erosive periods of the year. It is a surrogate for erosion, and along with land use intensity and diversity, it will give increased understanding on the issues of desertification.

A second set of core LQIs were recommended for longer-term research. These are indicators which require further development of their theoretical base, or lack adequate data for immediate development. These include:

- Soil Quality: Likely to be based on soil organic matter turn-over, particularly the dynamic (microbiological) carbon pool most affected by environmental conditions and land use change.
- Land degradation (erosion, salinization, compaction, organic matter loss): These processes have been much researched and have a strong scientific base, but reliable data on extent and impacts are often lacking.
- Agro-biodiversity: Involves objectives of managing natural habitats and the co-existence of native species in agricultural areas, maintaining natural soil micro and meso biodiversity, and managing the gene pools utilized in crop and animal production.

Four additional sets of core LQIs were identified, but these were recommended to be developed through collaboration with the respective authoritative disciplines:

- Water quality.
- Forest land quality.
- Rangeland quality.
- Land contamination/pollution.

The above are the biophysical components of sustainable land management. Although useful in their own right, they must still be complemented with indicators of the other pillars of sustainable land management, economic viability, system resilience, and social equity and acceptability. Considerable additional work is required to develop these pillars to the same level of detail as the land quality (biophysical) indicators.

Watershed Management

Watershed is defined as a geohydrological unit draining to a common point by a system of drains. All lands on earth are part of one watershed or other. Watershed is thus the land and water area, which contributes runoff to a common point.

A watershed is an area of land and water bounded by a drainage divide within which the surface runoff collects and flows out of the watershed through a single outlet into a lager river (or) lake.

Types of Watershed

Watersheds is classified depending upon the size, drainage, shape and land use pattern.

- Macro watershed (> 50,000 Hect).
- Sub-watershed (10,000 to 50,000 Hect).
- Milli-watershed (1000 to10000 Hect).
- Micro watershed (100 to 1000 Hect).
- Mini watershed (1-100 Hect).

Objectives of Watershed Management

The different objectives of watershed management programmes are:

- To control damaging runoff and degradation and thereby conservation of soil and water.
- To manage and utilize the runoff water for useful purpose.
- To protect, conserve and improve the land of watershed for more efficient and sustained production.
- To protect and enhance the water resource originating in the watershed.
- To check soil erosion and to reduce the effect of sediment yield on the watershed.
- To rehabilitate the deteriorating lands.
- To moderate the floods peaks at down stream areas.
- To increase infiltration of rainwater.
- To improve and increase the production of timbers, fodder and wild life resource.
- To enhance the ground water recharge, wherever applicable.

Factors affecting Watershed Management

Watershed Characters

- Size and shape,
- Topography,
- Soils,
- Relief.

Climatic Characteristic

- Precipitation,
- Amount and intensity of rainfall.

Watershed Operation

Land use Pattern:

- Vegetative cover,
- Density.

Social Status of Inhability

Water resource and their capabilities.

Watershed Management Practices:

- Interms of purpose.
- To increase infiltration.
- To increase water holding capacity.
- To prevent soil erosion.
- Method and accomplishment.

In brief various control measures are:

- Vegetative measures (Agronomical measures).
- Strip cropping.
- Pasture cropping.
- Grass land farming.
- Wood lands.
- Engineering measures (Structural practices o).
- Contour bunding.
- Terracing.
- Construction of earthern embankment.
- Construction of check dams.

- Construction of farm ponds.
- Construction of diversion.
- Gully controlling structure.
- Rock dam.
- Establishment of permanent grass and vegetation.
- Providing vegetative and stone barriers.
- Construction of silt tanks dentension.

Influence of soil conservation measures and vegetation cover on erosion, Runoff and Nutrient loss. Rainwater harvesting is the main component of watershed management. Some of the watershed management structures are as follows.

Broad Beds and Furrows

Function:

To control erosion and to conserve soil moisture in the soil during rainy days.

General Information:

The broad bed and furrow system is laid within the field boundaries. The land levels taken and it is laid using either animal drawn or tractor drawn ridgers.

Cost:

Approximate cost for laying beds & furrows is $1800/ha.

Salient Features:

- Conserves soil moisture in dryland.
- Controls soil erosion.
- Acts as a drainage channel during heavy rainy days.

Contour Bund

Function:

To intercept the run off flowing down the slope by an embankment.

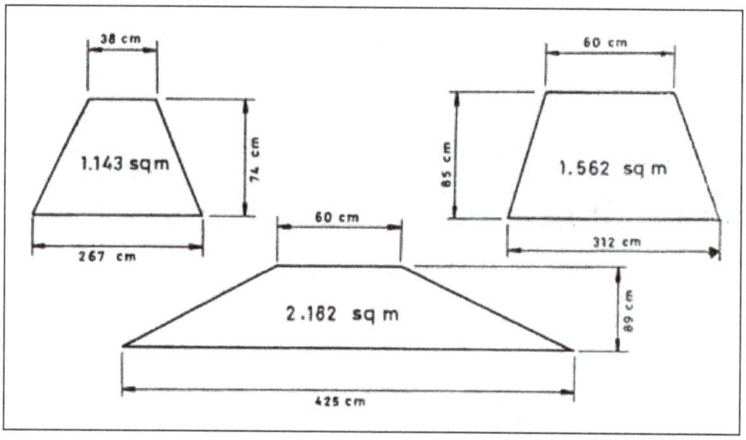

Contour bunding.

General Information:

It helps to control run off velocity. The embankment may be closed or open, surplus arrangements are provided wherever necessary.'

Cost:

Approximate cost of laying contour bund is $1400/ha.

Salient Features:

- It can be adopted on all soils.
- It can be laid upto 6% slopes.
- It helps to retain moisture in the field.

Bench Terracing

Function:

It helps to bring sloping land into different level strips to enable cultivation.

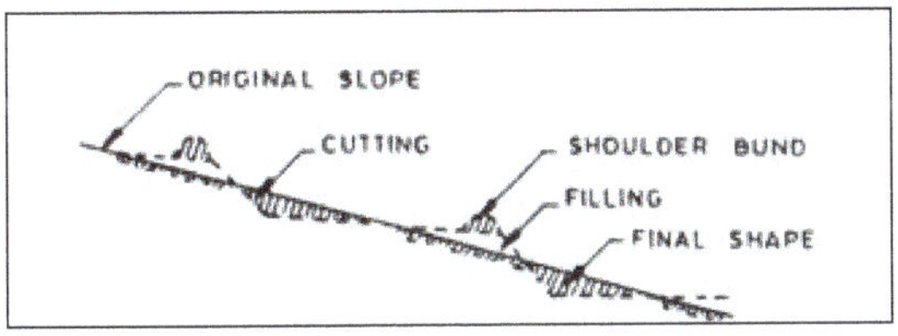

Level and table top.

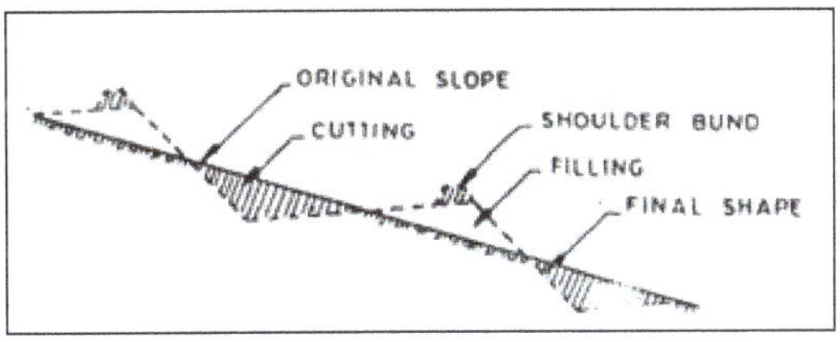

Sloping inward.

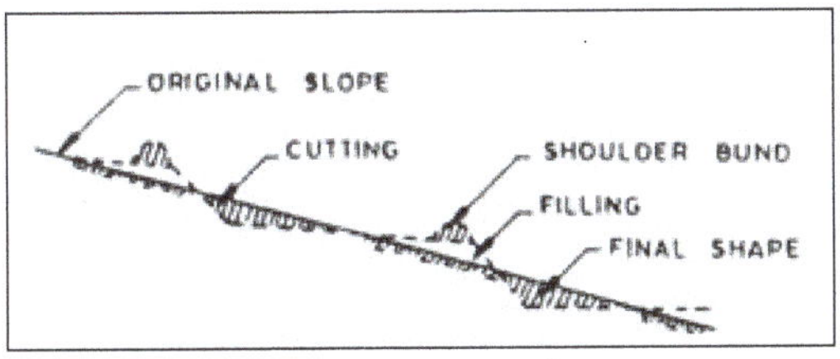

Sloping outward.

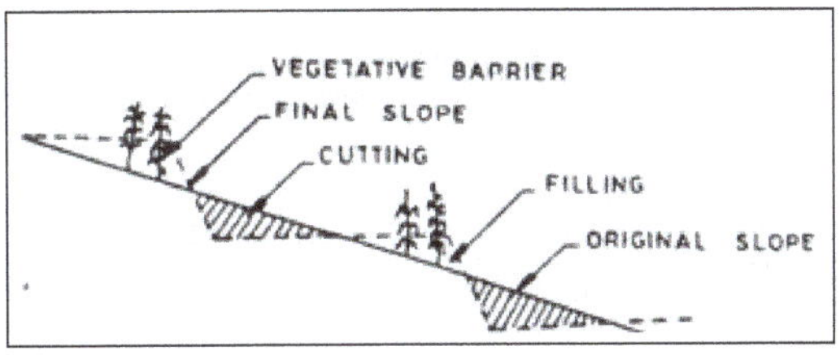

Puertorico type terrace.

Types of Bench Terraces

General Information:

It consists of construction of step like fields along contours by half cutting and half filling. Original slope is converted into level fields. The vertical & horizontal intervals are decided based on level slope.

Cost:

Approximate cost for laying the terrace is \$5000/ha.

Salient Features:

- Suitable for hilly regions.
- The benches may be inward sloping to drain off excess water.
- The outward sloping benches will help to reduce the existing steep slope to mild one.
- It is adopted in soils with slopes greater than 6%.

Microcatchments for Sloping Lands Function

It is useful for insitu moisture conservation and erosion control for tree crops.

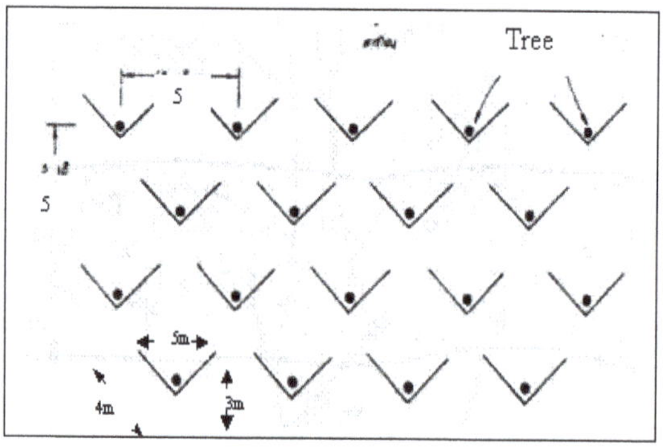

Microcatchments.

General Information:

Technique	Storage capacity per unit (m^3)	Annual run-off contribution to soil moisture Additional water stored per ha (m^3)	Estimated surface Run-off control (%)
Triangular catchments (V-Bunds)	13	5200	80
Crescent bunds	10.2	4080	73

Cost:

Technique	Cost/ha (Rs.)
Triangular catchments (V-Bunds)	6000-7000
Crescent bunds	2500-3000

Salient Features:

- Slope ranges from 2 –8%.
- Soil type – Light to moderate texture.

- Insitu moisture conservation with staggered planting.
- Suitable for dry land Horticulture & Agroforestry.
- Bund height – 30 to 45 cm.

Check Dam

Salient Features:

- A low weir normally constructed across the gullies.
- Constructed on small streams and long gullies formed by erosive activity of flood water.
- It cuts the velocity and reduces erosive activity.
- The stored water improves soil moisture of the adjoining area and allows percolation to recharge the aquifers.
- Spacing between the check dams water spread of one should be beyond the water spread of the other.
- Height depends on the bank height, varies from a metre to 3 metre and length varies from less than 3m to 10m.
- Cost varies from $ 40000/- to $ 100000/- per unit.

Percolation Pond

Function:

To augment the ground water recharge.

Salient Features:

- Shallow depression created at lower portions in a natural or diverted stream course.
- Preferable under gentle sloping stream where narrow valley exists.
- Located in soils of permeable nature.
- Adaptable where 20-30 ground water wells for irrigation exist with in the zone of influence about 800 – 900m.
- Minimum capacity may be around 5000 m^3 for the sack of economy.
- Also act as silt detention reservoir.
- Cost varies from $ 60000 to 150000 per unit.

Stone Barriers

Environmental Monitoring

Environmental monitoring is a tool to assess environmental conditions and trends, support policy development and its implementation, and develop information for reporting to national policymakers, international forums and the public.

Over the past decade, only a few countries of Europe and Central Asia have been able to maintain existing monitoring activities. The monitoring of urban air pollution – an important human health risk – is poor in many cities of the subregion. Solid and hazardous waste monitoring is weak and industrial emissions are also not well monitored, reducing the effectiveness of policy instruments such as emissions charges and fines. Monitoring of transboundary air pollution also needs strengthening. Moreover, many European and Central Asian countries lack uniform national methodologies across different monitoring areas, and their classification systems are often incompatible with international standards.

At the Fifth Environment for Europe Conference, UNECE Ministers endorsed recommendations for strengthening environmental monitoring and information systems in European and Central Asian countries prepared by the UNECE Working Group on Environmental Monitoring. Ministers also endorsed the UNECE guidelines for the preparation of national state-of-the environment reports. Together, these documents provide a road map for strengthening monitoring and reporting in the European and Central Asian subregion.

Air Monitoring

Air pollutants are known for their adverse effects on human health and ecosystems. Some of these pollutants also erode technical infrastructure and cultural monuments. Emissions of nitrogen oxides and non-methane volatile organic compounds are the main causes of the formation of ground-level ozone, which has adverse effects on human health and ecosystems. The air pollutants indicator assesses pressures from specific pollutants on atmospheric air across individual countries, but also identifies pressures from particular national sectors like energy, transport, industrial processes, agriculture and waste management.

On the basis of this indicator, public authorities can adjust the national environmental policy by, for instance, revising emission standards and emission limit values, strengthening permitting of potentially polluting activities and improving the application of economic instruments. Information on pollutant emissions is necessary for the assessment of transboundary air pollution and for international cooperation to address this problem.

Water Monitoring

Renewable freshwater resources have major environmental and economic value. Their distribution varies widely among and within countries. Pressures on freshwater resources are exerted by overexploitation and by pollution. Relating resources abstraction to renewal of stocks is a central issue in sustainable freshwater resource management. If a significant share of a country's water comes from transboundary rivers, tensions between countries can arise, especially if water availability in the upstream country is greater than in the downstream one. Countries are quite interdependent with regard to water resources.

The Convention on the Protection and Use of Transboundary Watercourses and International Lakes requires that the Parties introduce sustainable water management, including an ecosystem approach and the rational and fair use of transboundary waters.

Biodiversity

Sustainable development depends on a sound environment, which in turn depends on ecosystem diversity. Protected areas, especially the full range of International Union for Conservation of Nature (IUCN) Protected Area Categories, are essential for conserving biodiversity and contributing to sustainable development.

The biodiversity indicator provides a means to measure the response to the degradation of ecosystems and the loss of biodiversity in a country. It demonstrates the extent to which areas important for conserving biodiversity, cultural heritage, scientific research, recreation, natural resource maintenance and other environmental values are protected from incompatible uses.

Waste Monitoring

Waste represents a considerable loss of resources in the form of materials and energy. The treatment and disposal of waste may cause environmental pollution and expose humans to harmful substances and infectious organisms. Waste generation is closely linked to the level of economic activity in a country, and reflects the society's production and consumption patterns. A reduction in the volume of waste generated is therefore an indication of an economy's move towards less material-intensive production and consumption patterns.

The main purpose of the waste indicator is to measure the pressure on the environment of the total amount of generated waste and waste by category. The waste intensity represents a driving force indicator and shows response to anthropogenic activities.

Remote Sensing

Remote sensing can play an important role in monitoring and reporting on environmental problems, in particular when the target of such observations is to assess pollutant effects at large space scales over long time periods — i.e., at the regional, continental or even global scale over the entire seasonal cycle for a number of years.

Remote sensing can provide complementary information to existing ground-based environmental monitoring systems. It can be used to meet the need for timely information and can provide synoptic cross-boundary information. Data and information obtained through Earth observation can be used within geographic information systems for overlay and comparison with other geo-referenced information.

Enterprise Monitoring

Enterprise environmental monitoring and reporting is the system of measures implemented and paid for by operators, the natural or legal persons exercising actual power over the technical functioning of the facility. Such a system includes continuous and periodic observations, the recording, storage and treatment of data relating to environmental protection and the reporting of the results to the management and all the employees of enterprises, the public authorities and the general public as sets of primary, calculated or aggregated data and general information.

Strengthening enterprise environmental monitoring and reporting will improve monitoring of enterprise compliance with environmental regulations. Increasing the quantity of environmental information produced by enterprises, improving the quality of this information and enhancing access to it by the general public will help to exert significant pressure on polluters to reduce their adverse environmental impact.

Globalization and its Impact on the Environment

Globalization has had far-reaching effects on our lifestyle. It has led to faster access to technology, improved communication and innovation. Apart from playing an important role in bringing people of different cultures together, it has ushered a new era in the economic prosperity and has opened up vast channels of development. However, globalization has also created some areas of concern, and prominent among these is the impact that it has had on the environment. Globalization has featured extensively in the debates on environmentalism, and green activists have highlighted its far-reaching effects. Let us know about the impact of globalization on our environment.

Activists have pointed out that globalization has led to an increase in the consumption of products, which has impacted the ecological cycle. Increased consumption leads to an increase in the production of goods, which in turn puts stress on the environment. Globalization has also led to an increase in the transportation of raw materials and food from one place to another. Earlier, people used to consume locally-grown food, but with globalization, people consume products that have been developed in foreign countries. The amount of fuel that is consumed in transporting these products has led to an increase in the pollution levels in the environment. It has also led to several other environmental concerns such as noise pollution and landscape intrusion. Transportation has also put a strain on the non-renewable sources of energy, such as gasoline. The gases that are emitted from the aircraft have led to the depletion of the ozone layer apart from increasing the greenhouse effect. The industrial waste that is generated as a result of production has been laden on ships and dumped in oceans. This has killed many underwater organisms and has deposited many harmful chemicals in the ocean. The damage caused to ecosystem from the oil that spilled from one of the leaking containers of British Petroleum in 2010 is just one of the examples of the threat globalization poses to the environment.

Due to globalization and industrialization, various chemicals have been thrown into the soil which have resulted into the growth of many noxious weeds and plants. This toxic waste has caused a lot of damage to plants by interfering in their genetic makeup. It has put pressure on the available land resources. In various parts of the world, mountains are being cut to make way for a passing tunnel or a highway. Vast barren lands have been encroached upon to pave way for new buildings. While humans may rejoice on the glimmer with these innovations, these can have long-term effects on the environment. Various studies over the years, have found that plastic is one of the major toxic pollutants, as it is a non-biodegradable product. However, plastic is of immense use when it comes to packaging and preserving goods that are to be exported. This has led to increased use of plastic, causing widespread environmental pollution.

It has made so many changes in our lives that reversing it is not possible at all. The solution lies in developing effective mechanisms that can check the extent to which it can impact the environment. Researchers are of the view that the answer to this problem lies in the problem itself, that is, globalization itself can lend support to building a better structure which is economically feasible and environment-friendly. Globalization is about competition, and if certain privately owned companies can take the lead in being environment friendly, then it will encourage others to follow suit.

It is important that we put in some efforts to maintain harmony with the environment. The survival of human race on this planet is dependent on the environment to such a large extent that we cannot afford to ignore the consequences of our own actions. While there is a lot of debate and discussion on this issue, the need of the hour is to have effective policies in place, and implementation of those policies. The people that we have chosen to represent us have the responsibility of ensuring that the extent of damage on environment is curtailed, if not totally prevented.

Environmental Engineering

Environmental engineering is a professional engineering discipline that takes from broad scientific topics like chemistry, biology, ecology, geology, hydraulics, hydrology, microbiology, and mathematics to create solutions that will protect and also improves the health of living organisms and improve the quality of the environment. Environmental engineering is a sub-discipline of civil engineering and chemical engineering.

Environmental engineering is the application of scientific and engineering principles to improve and maintain the environment to:

- Protect human health,
- Protect nature's beneficial ecosystems,
- Improve environmental-related enhancement of the quality of human life.

Environmental engineers devise solutions for waste water management, water and air pollution control, recycling, waste disposal, and public health. They design municipal water supply and industrial wastewater treatment systems, and design plans to prevent waterborne diseases and improve sanitation in urban, rural and recreational areas. They evaluate hazardous-waste management systems to evaluate the severity of such hazards, advise on treatment and containment, and develop regulations to prevent mishaps. They implement environmental engineering law, as in assessing the environmental impact of proposed construction projects.

Environmental engineers study the effect of technological advances on the environment, addressing local and worldwide environmental issues such as acid rain, global warming, ozone depletion, water pollution and air pollution from automobile exhausts and industrial sources.

Most jurisdictions impose licensing and registration requirements for qualified environmental engineers.

Environmental engineering is a name for work that has been done since early civilizations, as people learned to modify and control the environmental conditions to meet needs. As people recognized that their health was related to the quality of their environment, they built systems to improve it. The ancient Indus Valley Civilization (3300 B.C.E. to 1300 B.C.E.) had advanced control over their water resources. The public work structures found at various sites in the area include wells, public baths,water storage tanks, a drinking water system, and a city-wide sewage collection system. They also had an early canal irrigation system enabling large-scale agriculture.

From 4000 to 2000 B.C.E., many civilizations had drainage systems and some had sanitation facilities, including the Mesopotamian Empire, Mohenjo-Daro, Egypt, Crete, and the Orkney Islands in Scotland. The Greeks also had aqueducts and sewer systems that used rain and wastewater to irrigate and fertilize fields.

The first aqueduct in Rome was constructed in 312 B.C.E., and from there, they continued to construct aqueducts for irrigation and safe urban water supply during droughts. They also built an underground sewer system as early as the 7th century B.C.E. that fed into the Tiber River, draining marshes to create farmland as well as removing sewage from the city.

Very little change was seen from the fall of Rome until the 19th century, where improvements saw increasing efforts focused on public health. Modern environmental engineering began in London in the mid-19th century when Joseph Bazalgette designed the first major sewerage system following the Great Stink. The city's sewer system conveyed raw sewage to the River Thames, which also supplied the majority of the city's drinking water, leading to an outbreak of cholera. The introduction of drinking water treatment and sewage treatment in industrialized countries reduced waterborne diseases from leading causes of death to rarities.

The field emerged as a separate academic discipline during the middle of the 20th century in response to widespread public concern about water and air pollution and other environmental degradation. As society and technology grew more complex, they increasingly produced unintended effects on the natural environment. One example is the widespread application of the pesticide DDT to control agricultural pests in the years following World War II. While the agricultural benefits were outstanding and crop yields increased dramatically, reducing world hunger, and malaria was controlled better than ever before, the pesticide brought numerous bird species to the edge of extinction due to its impact on their reproductive cycle. The story of DDT as vividly told in Rachel Carson's *Silent Spring* is considered to be the birth of the modern environmental movement, which led to the modern field of "environmental engineering."

Applications

Water Supply and Treatment

Environmental engineers evaluate the water balance within a watershed and determine the available water supply, the water needed for various needs in that watershed, the seasonal cycles of water movement through the watershed and they develop systems to store, treat, and convey water for various uses.

Water is treated to achieve water quality objectives for the end uses. In the case of a potable water supply, water is treated to minimize the risk of infectious disease transmission, the risk of non-infectious illness, and to create a palatable water flavor. Water distribution systems are designed and built to provide adequate water pressure and flow rates to meet various end-user needs such as domestic use, fire suppression, and irrigation.

Wastewater Treatment

Sewage treatment plant, Australia.

There are numerous wastewater treatment technologies. A wastewater treatment train can consist of a primary clarifier system to remove solid and floating materials, a secondary treatment system consisting of an aeration basin followed by flocculation and sedimentation or an activated sludge system and a secondary clarifier, a tertiary biological nitrogen removal system, and a final disinfection process. The aeration basin/activated sludge system removes organic material by growing bacteria (activated sludge). The secondary clarifier removes the activated sludge from the water. The tertiary system, although not always included due to costs, is becoming more prevalent to remove nitrogen and phosphorus and to disinfect the water before discharge to a surface water stream or ocean outfall.

Air Pollution Management

Scientists have developed air pollution dispersion models to evaluate the concentration of a pollutant at a receptor or the impact on overall air quality from vehicle exhausts and industrial flue gas stack emissions. To some extent, this field overlaps the desire to decrease carbon dioxide and other greenhouse gas emissions from combustion processes.

Environmental Impact Assessment and Mitigation

Water pollution.

Environmental engineers apply scientific and engineering principles to evaluate if there are likely to be any adverse impacts to water quality, air quality, habitat quality, flora and fauna, agricultural capacity, traffic, ecology, and noise. If impacts are expected, they then develop mitigation measures to limit or prevent such impacts. An example of a mitigation measure would be the creation of wetlands in a nearby location to mitigate the filling in of wetlands necessary for a road development if it is not possible to reroute the road.

In the United States, the practice of environmental assessment was formally initiated on January 1, 1970, the effective date of the National Environmental Policy Act (NEPA). Since that time, more than 100 developing and developed nations either have planned specific analogous laws or have adopted procedure used elsewhere. NEPA is applicable to all federal agencies in the United States.

Environmental Technology

Environmental technology, also known as 'green' or 'clean' technology, refers to the application of environmental sciences in the development of new technologies which aim to conserve, monitor or reduce the harm humans regularly cause the environment while consuming its resources. Sustainable development sits at the core of environmental technology - adopted practices that fuel economic development by avoiding the depletion of natural resources and further polluting.

Put simply, environmental technologies aim to protect the environment. They offer ways of consuming which are less polluting or do so in a sustainable manner, and often provide new ways to avoid depletion of natural resources altogether. Prominent examples include solar and wind energy, water desalination (the removal of salt or other minerals from saline water), electric vehicles, and pyrolysis (thermochemical decomposition of organic material).

Innovative emerging Environmental Technologies

Dozens of new and innovative environmental technologies emerge every year, some ready for mass consumption, and many more at prototype and proof of concept stages - all ultimately connected by the desire to provide sustainable tools for our collective use. Here are a few examples of recent innovations which have the potential to influence and shape our future environmental processes.

A recent revolutionary breakthrough in plastics manufacturing uses heliostats (appliances with driven mirrors to reflect light) to concentrate and reflect the sun's rays in order to melt plastic and make it moldable. Traditional plastic manufacturing uses fossil fuels as a feedstock of the product and to provide energy for the manufacturing process, but LightManufacturing's innovative new process is entirely fossil fuel-free. The process is not limited to new plastics and can be used to mold recycled plastics too.

Within the realm of recycling and wastewater treatment, a physicist from the National Taiwan University, Din Ping Tsai has developed a low-power method for treating wastewater using ultraviolet light and zinc oxide applied to CD-ROMs. Abundant by the millions, CDs are now seldom used, and offer a cheap and readily available source for wastewater treatment. During recent trials in which the spinning disks consume very little power, the device broke down over 95% of the water contaminants after an hour of treatment.

Environmental Monitoring

A significant part of environmental technology is environmental monitoring, which is the processes and activities performed in order to monitor the quality of the environment. Environmental monitoring has emerged as an essential component of governmental and private organisation policies across the globe.

Reference

- The-importance-of-environmental-protection-environmental-sciences-essay, environmental-sciences, essays: ukessays.com, Retrieved 6 January, 2020
- Environmental-management-concept-scope-and-aspects-of-environmental-management, environmental-management: environmentalpollution.in, Retrieved 13 May, 2020
- Environmental-management-its-characteristics-and-objectives, environment: environmentalpollution.in, Retrieved 8 January, 2020
- Approaches-to-environmental-management-aspects-and-types, environment: environmentalpollution.in, Retrieved 26 July, 2020
- Conservation-of-biodiversity, biodiversity-1, environment, energy: vikaspedia.in, Retrieved 19 April, 2020
- Agri-majorareas-watershed-watershedmgt, agriculture: tnau.ac.in, Retrieved 25 February, 2020
- "Careers in Environmental Engineering and Environmental Science". American Academy of Environmental Engineers & Scientists. Retrieved 2019-03-23
- Areas-of-work/environmental-monitoring, environmental-monitoring-and-assessment, environmental-policy: unece.org, Retrieved 31 March, 2020
- "Architecture and Engineering Occupations". Occupational Outlook Handbook. Bureau of Labor Statistics. 20 February 2019. Retrieved 23 March 2020
- What-is-environmental-technology, breaking-news, environmental-laboratory, news: envirotech-online.com, Retrieved 17 May, 2020

Key Concepts in Environmental Protection and Management 2

- Tragedy of the Commons
- Environmental Issue
- Environmental Psychology
- Corporate Environmental Responsibility
- Sustainability and Environmental Management
- Ecosystem Services
- Habitat Conservation
- Secondary Spill Containment
- Life-Cycle Assessment
- Waste Minimisation
- Environmental Management System

Environmental protection and management incorporates a number of concepts such as sustainable and environmental management, ecosystem services, habitat conservation, life-cycle assessment, waste minimisation and environmental management system, etc. This chapter closely examines these key concepts of environmental protection and management to provide an extensive understanding of the subject.

Tragedy of the Commons

Tragedy of the commons is the concept highlighting the conflict between individual and collective rationality.

The idea of the tragedy of the commons was made popular by the American ecologist Garrett Hardin, who used the analogy of ranchers grazing their animals on a common field. When the field is not over capacity, ranchers may graze their animals with few limitations. However, the rational rancher will seek to add livestock, thereby increasing profits. Thinking logically but not collectively, the benefits of adding animals adhere to the rancher alone, while the costs are shared. The tragedy is that ultimately no rancher will be able to graze the field, due to overconsumption. This scenario is played out on a daily basis in numerous instances, having grave consequences for the world's resources.

It is commonly recognized that one of the primary roles of government at the local, state, national, and international levels is to define and manage shared resources. However, there are a number of practical problems associated with this. Management inside clear political boundaries is a relatively straightforward task, but more problematic are resources shared across jurisdictions. For example, neighbouring cities may seek to maximize their benefits by competing for industry, but they may minimize their costs by pushing residents outside their jurisdictions. Another dimension is added at the international level when states are not bound by a common authority and may view restrictions on resource extraction as a threat to their sovereignty. Additional difficulties arise when resources cannot be divided or are interrelated, such as in whale hunting treaties when the fishing of the whales' food source is separately regulated.

The mechanisms to resolve these tragedies are part of a larger set of theories dealing with social dilemmas in fields such as mathematics, economics, sociology, urban planning, and environmental sciences. In these arenas, scholars have identified and structured a number of tentative solutions, such as enclosing the commons by establishing property rights, regulating through government intervention, or developing strategies to trigger collective behaviour. The American political scientist Elinor Ostrom, who was a cowinner of the 2009 Nobel Prize in Economic Sciences, argued that these strategies generally deal with problems of commitment and problems of mutual monitoring.

As the world's population rises and demands more access to resources, the issues associated with the commons become more severe. Ultimately, this may test the role and practicality of nation-states, leading to a redefinition of international governance. Among other important questions to consider is the proper role of supranational governments, such as the United Nations and the World Trade Organization. As resources become more limited, some argue, managing the commons may have neither a technical nor a political solution. This, indeed, may be the ultimate tragedy.

Environmental Issue

Environmental issues are harmful effects of human activity on the biophysical environment. Environmental protection is a practice of protecting the natural environment on individual, organizational or governmental levels, for the benefit of both the environment and humans. Environmentalism, a social and environmental movement, addresses environmental issues through advocacy, education and activism.

The carbon dioxide equivalent of greenhouse gases (GHG) in the atmosphere has already exceeded over 9000 parts per million (NOAA) (with total "long-term" GHG exceeding 455 parts per million) (Intergovernmental Panel on Climate Report). The amount of greenhouse gas in the atmosphere is possibly above the threshold that can potentially cause climate change. The UN Office for the Coordination of Humanitarian Affairs (OCHA) has stated "Climate change is not just a distant future threat. It is the main driver behind rising humanitarian needs and we are seeing its impact. The number of people affected and the damages inflicted by extreme weather has been unprecedented." Further, OCHA has stated:

> Climate disasters are on the rise. Around 70 percent of disasters are now climate related – up from around 50 percent from two decades ago.
>
> These disasters take a heavier human toll and come with a higher price tag. In the last decade, 2.4 billion people were affected by climate related disasters, compared to 1.7 billion in the previous decade. The cost of responding to disasters has risen tenfold between 1992 and 2008.
>
> Destructive sudden heavy rains, intense tropical storms, repeated flooding and droughts are likely to increase, as will the vulnerability of local communities in the absence of strong concerted action.

Environment destruction caused by humans is a global problem, and this is a problem that is on going every day. By year 2050, the global human population is expected to grow by 2 billion people, thereby reaching a level of 9.6 billion people. The human effects on Earth can be seen in many different ways. A main one is the temperature rise, and according to the report "Our Changing Climate", the global warming that has been going on for the past 50 years is primarily due to human activities. Since 1895, the U.S. average temperature has increased from 1.3 °F to 1.9 °F, with most of the increase taken place since around year 1970.

Types

Major current environmental issues may include climate change, pollution, environmental degradation, and resource depletion etc. The conservation movement lobbies for protection of endangered species and protection of any ecologically valuable natural areas, genetically modified foods and global warming.

Scientific Grounding

The level of understanding of Earth has increased markedly in recent times through science especially with the application of the scientific method. Environmental science is now a multi-disciplinary academic study taught and researched at many universities. This is used as a basis for addressing environmental issues.

Large amounts of data have been gathered and these are collated into reports, of which a common type is the State of the Environment publications. A recent major report was the Millennium Ecosystem Assessment, with input from 1200 scientists and released in 2005, which showed the high level of impact that humans are having on ecosystem services.

Solutions

Sustainability is the key to prevent or reduce the effect of environmental issues. There is now clear scientific evidence that humanity is living unsustainably, and that an unprecedented collective effort is needed to return human use of natural resources to within sustainable limits. For humans to live sustainably, the Earth's natural resources must be used at a rate at which they can be replenished (and by limiting global warming).

Concerns for the environment have prompted the formation of green parties, political parties that seek to address environmental issues. Initially these were formed in Australia, New Zealand and Germany but are now present in many other countries.

Environmental Degradation

Environmental degradation is the deterioration of the environment through depletion of resources such as air, water and soil; the destruction of ecosystems; habitat destruction; the extinction of wildlife; and pollution. It is defined as any change or disturbance to the environment perceived to be deleterious or undesirable. As indicated by the I=PAT equation, environmental impact (I) or degradation is caused by the combination of an already very large and increasing human population (P), continually increasing economic growth or per capita affluence (A), and the application of resource-depleting and polluting technology (T).

Environmental degradation is one of the ten threats officially cautioned by the High-level Panel on Threats, Challenges and Change of the United Nations. The United Nations International Strategy for Disaster Reduction defines environmental degradation as "the reduction of the capacity of the environment to meet social and ecological objectives, and needs". Environmental degradation comes in many types. When natural habitats are destroyed or natural resources are depleted, the environment is degraded. Efforts to counteract this problem include environmental protection and environmental resources management.

Biodiversity Loss

Scientists assert that human activity has pushed the earth into a sixth mass extinction event. The loss of biodiversity has been attributed in particular to human overpopulation, continued human population growth and overconsumption of natural resources by the world's wealthy. The *Global Assessment Report on Biodiversity and Ecosystem Services* published by IPBES in 2019 posits that roughly one million species of plants and animals face extinction from anthropogenic causes, such as expanding human land use for industrial agriculture and livestock rearing, along with overfishing.

Water Degradation

One major component of environmental degradation is the depletion of the resource of fresh water on Earth. Approximately only 2.5% of all of the water on Earth is fresh water, with the rest being salt water. 69% of fresh water is frozen in ice caps located on Antarctica and Greenland, so only 30% of the 2.5% of fresh water is available for consumption. Fresh water is an exceptionally important resource, since life on Earth is ultimately dependent on it. Water transports nutrients, minerals and chemicals within the biosphere to all forms of life, sustains both plants and animals, and moulds the surface of the Earth with transportation and deposition of materials.

The current top three uses of fresh water account for 95% of its consumption; approximately 85% is used for irrigation of farmland, golf courses, and parks, 6% is used for domestic purposes such as indoor bathing uses and outdoor garden and lawn use, and 4% is used for industrial purposes such as processing, washing, and cooling in manufacturing centres. It is estimated that one in three people over the entire globe are already facing water shortages, almost one-fifth of the world population live in areas of physical water scarcity, and almost one quarter of the world's population live in a developing country that lacks the necessary infrastructure to use water from available rivers and aquifers. Water scarcity is an increasing problem due to many foreseen issues in the future including population growth, increased urbanization, higher standards of living, and climate change.

Climate Change and Temperature

Climate change affects the Earth's water supply in a large number of ways. It is predicted that the mean global temperature will rise in the coming years due to a number of forces affecting the climate. The amount of atmospheric carbon dioxide (CO_2) will rise, and both of these will influence water resources; evaporation depends strongly on temperature and moisture availability which can ultimately affect the amount of water available to replenish groundwater supplies.

Transpiration from plants can be affected by a rise in atmospheric CO_2, which can decrease their use of water, but can also raise their use of water from possible increases of leaf area. Temperature rise can reduce the snow season in the winter and increase the intensity of the melting snow leading to peak runoff of this, affecting soil moisture, flood and drought risks, and storage capacities depending on the area.

Warmer winter temperatures cause a decrease in snowpack, which can result in diminished water resources during summer. This is especially important at mid-latitudes and in mountain regions that depend on glacial runoff to replenish their river systems and groundwater supplies, making these areas increasingly vulnerable to water shortages over time; an increase in temperature will initially result in a rapid rise in water melting from glaciers in the summer, followed by a retreat in glaciers and a decrease in the melt and consequently the water supply every year as the size of these glaciers get smaller and smaller.

Thermal expansion of water and increased melting of oceanic glaciers from an increase in temperature gives way to a rise in sea level. This can affect the fresh water supply to coastal areas as well. As river mouths and deltas with higher salinity get pushed further inland, an intrusion of saltwater results in an increase of salinity in reservoirs and aquifers. Sea-level rise may also consequently be caused by a depletion of groundwater, as climate change can affect the hydrologic cycle in a number of ways. Uneven distributions of increased temperatures and increased precipitation around the globe results in water surpluses and deficits, but a global decrease in groundwater suggests a rise in sea level, even after meltwater and thermal expansion were accounted for, which can provide a positive feedback to the problems sea-level rise causes to fresh-water supply.

A rise in air temperature results in a rise in water temperature, which is also very significant in water degradation as the water would become more susceptible to bacterial growth. An increase in water temperature can also affect ecosystems greatly because of a species' sensitivity to temperature, and also by inducing changes in a body of water's self-purification system from decreased amounts of dissolved oxygen in the water due to rises in temperature.

Climate Change and Precipitation

A rise in global temperatures is also predicted to correlate with an increase in global precipitation but because of increased runoff, floods, increased rates of soil erosion, and mass movement of land, a decline in water quality is probable, because while water will carry more nutrients it will also carry more contaminants. While most of the attention about climate change is directed towards global warming and greenhouse effect, some of the most severe effects of climate change are likely to be from changes in precipitation, evapotranspiration, runoff, and soil moisture. It is generally expected that, on average, global precipitation will increase, with some areas receiving increases and some decreases.

Climate models show that while some regions should expect an increase in precipitation, such as in the tropics and higher latitudes, other areas are expected to see a decrease, such as in the subtropics. This will ultimately cause a latitudinal variation in water distribution. The areas receiving more precipitation are also expected to receive this increase during their winter and actually become drier during their summer, creating even more of a variation of precipitation distribution. Naturally, the distribution of precipitation across the planet is very uneven, causing constant variations in water availability in respective locations.

Changes in precipitation affect the timing and magnitude of floods and droughts, shift runoff processes, and alter groundwater recharge rates. Vegetation patterns and growth rates will be directly affected by shifts in precipitation amount and distribution, which will in turn affect agriculture as well as natural ecosystems. Decreased precipitation will deprive areas of water causing water tables to fall and reservoirs of wetlands, rivers, and lakes to empty. In addition, a possible increase in evaporation and evapotranspiration will result, depending on the accompanied rise in temperature. Groundwater reserves will be depleted, and the remaining water has a greater chance of being of poor quality from saline or contaminants on the land surface.

Population Growth

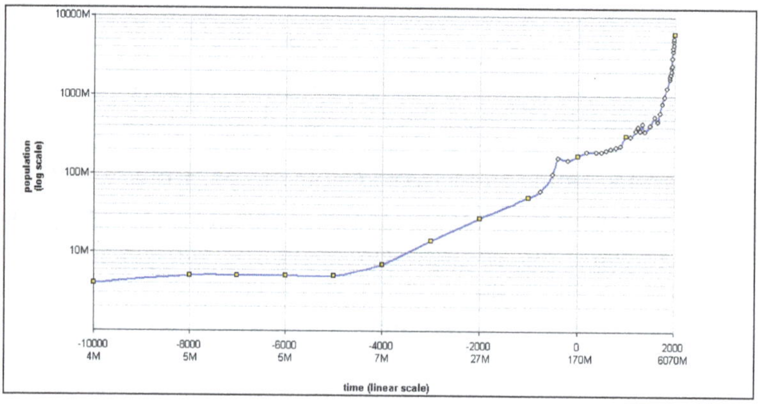

World population growth in a lin-log scale.

The human population on Earth is expanding rapidly which goes hand in hand with the degradation of the environment at large measures. Humanity's appetite for needs is disarranging the environment's natural equilibrium. Production industries are venting smoke and discharging chemicals that are polluting water resources. The smoke that is emitted into the atmosphere holds detrimental gases such as carbon monoxide and sulphur dioxide. The high levels of pollution in the atmosphere form layers that are eventually absorbed into the atmosphere. Organic compounds such as chlorofluorocarbons (CFC's) have generated an unwanted opening in the ozone layer, which emits higher levels of ultraviolet radiation putting the globe at large threat.

The available fresh water being affected by the climate is also being stretched across an ever-increasing global population. It is estimated that almost a quarter of the global population is living in an area that is using more than 20% of their renewable water supply; water use will rise with population while the water supply is also being aggravated by decreases in streamflow and groundwater caused by climate change. Even though some areas may see an increase in freshwater supply from an uneven distribution of precipitation increase, an increased use of water supply is expected.

An increased population means increased withdrawals from the water supply for domestic, agricultural, and industrial uses, the largest of these being agriculture, believed to be the major non-climate driver of environmental change and water deterioration. The next 50 years will likely be the last period of rapid agricultural expansion, but the larger and wealthier population over this time will demand more agriculture.

Population increase over the last two decades, at least in the United States, has also been accompanied by a shift to an increase in urban areas from rural areas, which concentrates the demand for water into certain areas, and puts stress on the fresh water supply from industrial and human contaminants. Urbanization causes overcrowding and increasingly unsanitary living conditions, especially in developing countries, which in turn exposes an increasingly number of people to disease. About 79% of the world's population is in developing countries, which lack access to sanitary water and sewer systems, giving rises to disease and deaths from contaminated water and increased numbers of disease-carrying insects.

Agriculture

Water pollution due to dairy farming in the Wairarapa.

Agriculture is dependent on available soil moisture, which is directly affected by climate dynamics, with precipitation being the input in this system and various processes being the output, such as evapotranspiration, surface runoff, drainage, and percolation into groundwater. Changes in climate, especially the changes in precipitation and evapotranspiration predicted by climate models, will directly affect soil moisture, surface runoff, and groundwater recharge.

In areas with decreasing precipitation as predicted by the climate models, soil moisture may be substantially reduced. With this in mind, agriculture in most areas already needs irrigation, which depletes fresh water supplies both by the physical use of the water and the degradation agriculture causes to the water. Irrigation increases salt and nutrient content in areas that would not normally be affected, and damages streams and rivers from damming and removal of water. Fertilizer enters both human and livestock waste streams that eventually enter groundwater, while nitrogen, phosphorus, and other chemicals from fertilizer can acidify both soils and water. Certain agricultural demands may increase more than others with an increasingly wealthier global population, and meat is one commodity expected to double global food demand by 2050, which directly affects the global supply of fresh water. Cows need water to drink, more if the temperature is high and humidity is low, and more if the production system the cow is in is extensive, since finding food takes more effort. Water is needed in processing of the meat, and also in the production of feed for the livestock. Manure can contaminate bodies of freshwater, and slaughterhouses, depending on how well they are managed, contribute waste such as blood, fat, hair, and other bodily contents to supplies of fresh water.

The transfer of water from agricultural to urban and suburban use raises concerns about agricultural sustainability, rural socioeconomic decline, food security, an increased carbon footprint from imported food, and decreased foreign trade balance. The depletion of fresh water, as applied to more specific and populated areas, increases fresh water scarcity among the population and also makes populations susceptible to economic, social, and political conflict in a number of ways; rising sea levels forces migration from coastal areas to other areas farther inland, pushing populations closer together breaching borders and other geographical patterns, and agricultural surpluses and deficits from the availability of water induce trade problems and economies of certain areas. Climate change is an important cause of involuntary migration and forced displacement According to the Food and Agriculture Organization of the United Nations, global greenhouse gas emissions from animal agriculture exceeds that of transportation.

Water Management

A stream in the town of Amlwch, Anglesey which is contaminated by acid mine drainage from the former copper mine at nearby Parys Mountain.

The issue of the depletion of fresh water can be met by increased efforts in water management. While water management systems are often flexible, adaptation to new hydrologic conditions may be very costly. Preventative approaches are necessary to avoid high costs of inefficiency and the need for rehabilitation of water supplies, and innovations to decrease overall demand may be important in planning water sustainability.

Water supply systems, as they exist now, were based on the assumptions of the current climate, and built to accommodate existing river flows and flood frequencies. Reservoirs are operated based on past hydrologic records, and irrigation systems on historical temperature, water availability, and crop water requirements; these may not be a reliable guide to the future. Re-examining engineering designs, operations, optimizations, and planning, as well as re-evaluating legal, technical, and economic approaches to manage water resources are very important for the future of water management in response to water degradation. Another approach is water privatization; despite its economic and cultural effects, service quality and overall quality of the water can be more easily controlled and distributed. Rationality and sustainability is appropriate, and requires limits to overexploitation and pollution and efforts in conservation.

Environmental Psychology

Environmental psychology is an interdisciplinary field that focuses on the interplay between individuals and their surroundings. It examines the way in which the natural environment and our built environments shape us as individuals. The field defines the term environment broadly, encompassing natural environments, social settings, built environments, learning environments, and informational environments.

Environmental psychology was not fully recognized as its own field until the late 1960s when scientists began to question the tie between human behavior and our natural and built environments. Since its conception, the field has been committed to the development of a discipline that is both value oriented and problem oriented, prioritizing research aimed at solving complex environmental problems in the pursuit of individual well-being within a larger society. When solving problems involving human-environment interactions, whether global or local, one must have a model of human nature that predicts the environmental conditions under which humans will respond well. This model can help design, manage, protect and restore environments that enhance reasonable behavior, predict the likely outcomes when these conditions are not met, and diagnose problem situations. The field develops such a model of human nature while retaining a broad and inherently multidisciplinary focus. It explores such dissimilar issues as common property resource management, wayfinding in complex settings, the effect of environmental stress on human performance, the characteristics of restorative environments, human information processing, and the promotion of durable conservation behavior. Lately, alongside the increased focus on climate change in society and the social sciences and the re-emergence of limits-to-growth concerns, there has been increased focus on environmental sustainability issues within the field.

This multidisciplinary paradigm has not only characterized the dynamic for which environmental psychology is expected to develop. It has also been the catalyst in attracting other schools of knowledge in its pursuit, aside from research psychologists. Geographers, economists, landscape architects, policy-makers, sociologists, anthropologists, educators, and product developers all have discovered and participated in this field.

Although "environmental psychology" is arguably the best-known and most comprehensive description of the field, it is also known as human Factors science, cognitive ergonomics, ecological psychology, ecopsychology, environment–behavior studies, and person–environment studies. Closely related fields include architectural psychology, socio-architecture, behavioral geography, environmental sociology, social ecology, and environmental design research.

The origins of this field of study are unknown, however, Willy Hellpach is said to be the first to mention "environmental psychology". One of his books, *Geopsyche*, discusses topics such as how the sun and the moon affect human activity, the impact of extreme environments, and the effects of color and form. Among the other major scholars at the roots of environmental psychology were Jakob von Uexküll, Kurt Lewin, Egon Brunswik, and later Gerhard Kaminski and Carl Friedrich Graumann.

The end of World War II brought about a higher demand for developments in the field of social psychology particularly in the areas of attitude change, small group processes, and intergroup conflict. This demand caused psychologists to begin applying social psychology theories to a number of social issues such as prejudice, war and peace. It was thought that if these problems were addressed, underlying notions and principles would surface.

Although this period was crucial to the development of the field, the methodologies used to carry out the studies were questionable. At the time, studies were being conducted in a laboratory setting, which caused some doubt as to their validity in the real world. Consequently, environmental psychologists began to conduct studies outside of the laboratory, enabling the field to continue to progress. Today environmental psychology is being applied to many different areas such as architecture and design, television programs and advertisements.

Orientations

Problem Oriented

Environmental psychology is a direct study of the relationship between an environment and how that environment affects its inhabitants. Specific aspects of this field work by identifying a problem and through the identification of said problem, discovering a solution. Therefore, it is necessary for environmental psychology to be problem oriented.

One important aspect of a problem-oriented field is that by identifying problems, solutions arise from the research acquired. The solutions can aid in making society function better as a whole and create a wealth of knowledge about the inner workings of societies. Environmental psychologist Harold Proshansky discusses how the field is also "value oriented" because of the field's commitment to bettering society through problem identification. Panyang discusses the importance of not only understanding the problem but also the necessity of a solution. Proshansky also points out some of the problems of a problem-oriented approach for environmental psychology. First the problems being identified must be studied under certain specifications: it must be ongoing and occurring in real life, not in a laboratory. Second, the notions about the problems must derive directly from the source – meaning they must come directly from the specific environment where the problem is occurring. The solutions and understanding of the problems cannot come from an environment that has been constructed and modeled to look like real life. Environmental psychology needs to reflect the actual society not a society built in a laboratory setting. The difficult task of the environmental psychologist is to study problems as they are occurring in everyday life. It is hard to reject all laboratory research because laboratory experiments are where theories may be tested without damaging the actual environment or can serve as models when testing solutions. Proshansky makes this point as well, discussing the difficulty in the overall problem oriented approach. He states that it is important, however, for the environmental psychologist to utilize all aspects of research and analysis of the findings and to take into account both the general and individualized aspects of the problems.

Environmental psychology addresses environmental problems such as density and crowding, noise pollution, sub-standard living, and urban decay. Noise increases environmental stress. Although it has been found that control and predictability are the greatest factors in stressful effects of noise; context, pitch, source and habituation are also important variables. Environmental psychologists have theorized that density and crowding can also have an adverse effect on mood and may cause stress-related illness. To understand and solve environmental problems, environmental psychologists believe concepts and principles should come directly from the physical settings and problems being looked at. For example, factors that reduce feelings of crowding within buildings include:

- Windows – particularly ones that can be opened and ones that provide a view as well as light.
- High ceilings.
- Doors to divide spaces (Baum and Davies)and provide access control.
- Room shape – square rooms feel less crowded than rectangular ones (Dresor).
- Using partitions to create smaller, personalized spaces within an open plan office or larger work space.

- Providing increases in cognitive control over aspects of the internal environment, such as ventilation, light, privacy, etc.
- Conducting a cognitive appraisal of an environment and feelings of crowding in different settings. For example, one might be comfortable with crowding at a concert but not in school corridors.
- Creating a defensible space (Calhoun).

Personal Space and Territory

Having an area of personal territory in a public space, e.g., at the office, is a key feature of many architectural designs. Having such a 'defensible space' can reduce the negative effects of crowding in urban environments. The term, coined by John B. Calhoun in 1947, is the result of multiple environmental experiments conducted on rats. Originally beginning as an experiment to measure how many rats could be accommodated in a given space, it expanded into determining how rats, given the proper food, shelter and bedding would behave under a confined environment.

Under these circumstances, the males became aggressive, some exclusively homosexual. Others became pansexual and hypersexual, seeking every chance to mount any rat they encountered. As a result, mating behaviors were upset with an increase in infant mortalities. With parents failing to provide proper nests, thoughtlessly ditching their young and even attacking them, infant mortality rose as high as 96% in certain sections.

Creating barriers and customizing the space are ways of creating personal space, e.g., using pictures of one's family in an office setting. This increases cognitive control as one sees oneself as having control over the competitors to the personal space and therefore able to control the level of density and crowding in the space.

Systems Oriented

The systems oriented approach to experimenting is applied to individuals or people that are a part of communities, groups, and organizations. This approach particularly examines group interaction, as opposed to an individual's interaction and it emphasizes on factors of social integration. In the laboratory, experiments focus on cause and effect processes within human nature.

Interdisciplinary Oriented

Environmental psychology relies on interaction with other disciplines in order to approach problems with multiple perspectives. The first discipline is the category of behavioral sciences, which include: sociology, political science, anthropology, and economics. Environmental psychology also interacts with the interspecializations of the field of psychology, which include: developmental psychology, cognitive science, industrial and organizational psychology, psychobiology, psychoanalysis, and social neuroscience. In addition to the more scientific fields of study, environmental psychology also works with the design field which includes: the studies of architecture, interior design, urban planning, industrial and object design, landscape architecture, and preservation.

Space-over-time Orientation

Space over time orientation highlights the importance of the past. Examining problems with the past in mind creates a better understanding of how past forces, such as social, political, and economic forces, may be of relevance to present and future problems. Time and place are also important to consider. It's important to look at time over extended periods. Physical settings change over time; they change with respect to physical properties and they change because individuals using the space change over time. Looking at these spaces over time will help monitor the changes and possibly predict future problems.

There are a variety of tests that can be administered to children in order to determine their temperament. Temperament is split up into three types: "easy", "difficult", and "slow-to-warm-up". Alexander Thomas, Stella Chess, Herbert G. Birch, Margaret Hertzig and Sam Korn created an infant temperament test in the 1950s and rated them using nine temperament criteria. By finding out a child's temperament at birth, it enables us to know what to expect as the child progresses into adulthood.

Place Identity

For many years Harold Proshansky and his colleagues at the Graduate School and University Center of the City University of New York, explored the concept of place identity. Place identity has been traditionally defined as a 'sub-structure of the self-identity of the person consisting of broadly conceived cognitions about the physical world in which the individual lives'. These cognitions define the daily experiences of every human being. Through one's attitudes, feelings, ideas, memories, personal values and preferences toward the range and type of physical settings, he/she can then understand the environment they live in and their overall experience.

As a person interacts with various places and spaces, he/she is able to evaluate which properties in different environments fulfill his/her various needs. When a place contains components that satisfy a person biologically, socially, psychologically and/or culturally, it creates the environmental past of a person. Through 'good' or 'bad' experiences with a place, a person is then able to reflect and define their personal values, attitudes, feelings and beliefs about the physical world.

Place identity has been described as the individual's incorporation of place into the larger concept of self; a "potpourri of memories, conceptions, interpretations, ideas, and related feelings about specific physical settings, as well as types of settings". Other theorists have been instrumental in the creation of the idea of place identity. Three humanistic geographers, Tuan, Relph and Buttimer, share a couple of basic assumptions. As a person lives and creates memories within a place, attachment is built and it is through one's personal connection to a place, that he/she gains a sense of belonging and purpose, which then gives significance and meaning to their life.

Five central functions of place-identity have been depicted: recognition, meaning, expressive-requirement, mediating change, and anxiety and defense function. Place identity becomes a cognitive "database" against which every physical setting is experienced. The activities of a person often overlap with physical settings, which then create a background for the rest of life's interactions and events. The individual is frequently unaware of the array of feelings, values or memories of a singular place and simply becomes more comfortable or uncomfortable with certain broad kinds of physical settings, or prefers specific spaces to others. In the time since the term "place identity" was introduced, the theory has been the model for identity that has dominated environmental psychology.

Place Attachment

Many different perceptions of the bond between people and places have been hypothesized and studied. The most widespread terms include place attachment and sense of place. One consistent thread woven throughout most recent research on place attachment deals with the importance of the amount of time spent at a certain place (the length of association with a place). While both researchers and writers have made the case that time and experience in a place are important for deepening the meanings and emotional ties central to the person-place relationship, little in-depth research has studied these factors and their role in forging this connection.

Place attachment is defined as one's emotional or affective ties to a place, and is generally thought to be the result of a long-term connection with a certain environment. This is different from a simple aesthetic response such as saying a certain place is special because it is beautiful. For example, one can have an emotional response to a beautiful (or ugly) landscape or place, but this response may sometimes be shallow and fleeting. This distinction is one that Schroeder labeled "meaning versus preference". According to Schroeder the definition of "meaning" is "the thoughts, feelings, memories and interpretations evoked by a landscape"; whereas "preference" is "the degree of liking for one landscape compared to another". For a deeper and lasting emotional attachment to develop (Or in Schroeder's terms, for it to have meaning) an enduring relationship with a place is usually a critical factor. Chigbu carried out a rural study of place-attachment using a qualitative approach to check its impact on a community, Uturu (in Nigeria), and found that it has direct relationship to level of community development.

Environmental Consciousness

Leanne Rivlin theorized that one way to examine an individual's environmental consciousness is to recognize how the physical place is significant, and look at the people/place relationship.

Environmental cognition (involved in human cognition) plays a crucial role in environmental perception. All different areas of the brain engage with environmentally relevant information. Some believe that the orbitofrontal cortex integrates environmentally relevant information from many distributed areas of the brain. Due to its anterior location within the frontal cortex, the orbitofrontal cortex may make judgments about the environment, and refine the organism's "understanding" through error analysis, and other processes specific to prefrontal cortex. But to be certain, there is no single brain area dedicated to the organism's interactions with its environment. Rather, all brain areas are dedicated to this task. One area (probably the orbitofrontal cortex) may collate the various pieces of the informational puzzle in order to develop a long term strategy of engagement with the ever-changing "environment." Moreover, the orbitofrontal cortex may show the greatest change in blood oxygenation (BOLD level) when an organism thinks of the broad, and amorphous category referred to as "the environment." Because of the recent concern with the environment, environmental consciousness or awareness has come to be related to the growth and development of understanding and consciousness toward the biophysical environment and its problems.

Behavior Settings

The earliest noteworthy discoveries in the field of environmental psychology can be dated back to Roger Barker who created the field of ecological psychology. Founding his research station in Oskaloosa, Kansas in 1947, his field observations expanded into the theory that social settings influence behavior. Empirical data gathered in Oskaloosa from 1947 to 1972 helped him develop the concept of the "behavior setting" to help explain the relationship between the individual and the immediate environment. This was further explored in his work with Paul Gump in the book *Big School, Small School: High School Size and Student Behavior*. One of the first insightful explanations on why groups tend to be less satisfying for their members as they increase in size, their studies illustrated that large schools had a similar number of behavior settings to that of small schools. This resulted in the students' ability to presume many different roles in small schools (e.g. be in the school band and the school football team) but in larger schools there was a propensity to deliberate over their social choices.

In his book *Ecological Psychology* Barker stresses the importance of the town's behavior and environment as the residents' most ordinary instrument of describing their environment. "The hybrid, eco-behavioral character of behavior settings appear to present Midwest's inhabitants with no difficulty; nouns that combine milieu and standing behavior are common, e.g. oyster supper, basketball game, turkey dinner, golden gavel ceremony, cake walk, back surgery, gift exchange, livestock auction, auto repair."

Barker argued that his students should implement T-methods (psychologist as 'transducer': i.e. methods in which they studied man in his 'natural environment') rather than O-methods (psychologist as "operators" i.e. experimental methods). Basically, Barker preferred fieldwork and direct observation rather than controlled experiments. Some of the minute-by-minute observations of Kansan children from morning to night, jotted down by young and maternal graduate students, may be the most intimate and poignant documents in social science. Barker spent his career expanding on what he called ecological psychology, identifying these behavior settings, and publishing accounts such as *One Boy's Day* and *Midwest and Its Children*.

Applications

Impact on the Built Environment

Environmental psychologists rejected the laboratory-experimental paradigm because of its simplification and skewed view of the cause-and-effect relationships of human's behaviors and experiences. Environmental psychologists examine how one or more parameters produce an effect while other measures are controlled. It is impossible to manipulate real-world settings in a laboratory.

Environmental psychology is oriented towards influencing the work of design professionals (architects, engineers, interior designers, urban planners, etc.) and thereby improving the human environment.

On a civic scale, efforts towards improving pedestrian landscapes have paid off, to some extent, from the involvement of figures like Jane Jacobs and Copenhagen's Jan Gehl. One prime figure here is the late writer and researcher William H. Whyte. His still-refreshing and perceptive "City", based on his accumulated observations of skilled Manhattan pedestrians, provides steps and patterns of use in urban plazas.

The role and impact of architecture on human behavior is debated within the architectural profession. Views range from: supposing that people will adapt to new architectures and city forms; believing that architects cannot predict the impact of buildings on humans and therefore should base decisions on other factors; to those who undertake detailed precedent studies of local building types and how they are used by that society.

Environmental psychology has conquered the whole architectural genre which is concerned with retail stores and any other commercial venues that have the power to manipulate the mood and behavior of customers (e.g. stadiums, casinos, malls, and now airports). From Philip Kotler's landmark paper on Atmospherics and Alan Hirsch's "Effects of Ambient Odors on Slot-Machine Usage in a Las Vegas Casino", through the creation and management of the Gruen transfer, retail relies heavily on psychology, original research, focus groups, and direct observation. One of William Whyte's students, Paco Underhill, makes a living as a "shopping anthropologist". Most of this advanced research remains a trade secret and proprietary.

Challenges

The field saw significant research findings and a fair surge of interest in the late 1970s and early 1980s, but has seen challenges of nomenclature, obtaining objective and repeatable results, scope, and the fact that some research rests on underlying assumptions about human perception, which is not fully understood. Being an interdisciplinary field is difficult because it lacks a solid definition and purpose. It is hard for the field to fit into organizational structures. In the words of Guido Francescato, speaking in 2000, environmental psychology encompasses a "somewhat bewildering array of disparate methodologies, conceptual orientations, and interpretations making it difficult to delineate, with any degree of precision, just what the field is all about and what might it contribute to the construction of society and the unfolding of history."

Environmental psychology has not received nearly enough supporters to be considered an interdisciplinary field within psychology. Harold M. Proshanksy was one of the founders of environmental psychology and was quoted as saying "As I look at the field of environmental psychology today, I am concerned about its future. It has not, since its emergence in the early 1960s grown to the point where it can match the fields of social, personality, learning or cognitive psychology. To be sure, it has increased in membership, in the number of journals devoted to it, and even in the amount of professional organizational support it enjoys, but not enough so that one could look at any major university and find it to be a field of specialization in a department of psychology, or, more importantly, in an interdisciplinary center or institute".

Corporate Environmental Responsibility

Corporate Environmental Responsibility (CER) refers to a company's duties to abstain from damaging natural environments. The term derives from corporate social responsibility (CSR).

The environmental aspect of CSR has been debated over the past few decades, as stakeholders increasingly require organizations to become more environmentally aware and socially responsible. In the traditional business model, environmental protection was considered only in relation to the "public interest". Hitherto, governments had maintained principal responsibility for ensuring environmental management and conservation.

The public sector has been focused on the development of regulations and the imposition of sanctions as a means to facilitating environmental protection. Recently, the private sector has adopted the approach of co-responsibility towards the prevention and alleviation of environmental damage. The sectors and their roles have been chang ing, with the private sector becoming more active in the protection of the environment. Many governments, corporations, and big companies are now providing strategies for environmental protection and economic growth.

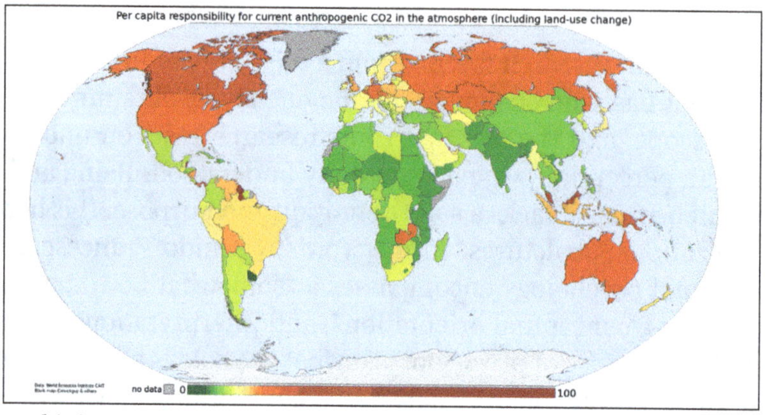

This image is a map of the global carbon emissions from 1950-2000.

The World Commission on Environment published the Brundtland Report in 1987 to address sustainable development. Since then, managers, scholars, and business owners have tried to determine why and how big corporations should incorporate environmental aspects into their own policies. In recent years, an increasing number of companies have pledged to protect natural environments.

Relations to Corporate Social Responsibility

There are different perceptions of CSR between government, the private sector, non-governmental organizations (NGOs) and society in general, and thus, the concept has no single definition.

CSR may cover:

- A company running its business responsibly in relation to internal stakeholders (shareholders, employees, customers and suppliers);
- The role of business in relation to the state (locally and nationally) as well as to inter-state institutions or standards; and
- Business performance as a responsible member of the society in which it operates and the global community."

The European Union defines CSR as "the concept that an enterprise is accountable for its impact on all relevant stakeholders. It is the continuing commitment by a business to behave fairly and responsibly and contribute to economic development while improving the quality of life of the workforce and their families as well as of the local community and society at large." According to this definition, a CSR strategy is more focused on social aspects, particularly the interests of stakeholders.

CER is, in many ways, connected to CSR, as both of them influence environmental protection. CER, however, is strictly about the consideration of environmental implications and protection within corporate strategy. The understanding of CER cannot be separated from CSR—both are interconnected and based on environmental protection. There are three major areas related to these two concepts—economic, environmental and social. CER is focused more on economic and environmental while CSR relates to social and environmental aspects. Economy, society, and environment all play significant roles in the development of an efficient and effective company strategy.

Main Elements

These cover the environmental implications of a company's operations:

- Eliminate waste and emissions.
- Maximize the efficient use of resources and productivity.
- Minimize activities that might impair the enjoyment of resources by future generations.

Drivers and Challenges

Among the main drivers for CER are government policies and regulations. Many states provide their own legislation, regulations and policies, which are important in creating a positive environmental attitude within companies. Subsidies, tariffs and taxes play a vital role in the implementation of these policies. Another significant factor is the competitive environment among companies generated by media, public, shareholder and NGO awareness, which are also major drivers of CER.

Challenges include the cost of regulation and difficulties in predicting economic gains, which could become problematic for a company's management. Additionally, new technologies are frequently too expensive for a lot of companies. Another challenge is the lack of harmonization of regulations among different states—often there is a mosaic of propositions, leading to unclear strategies for environmental behavior, especially in multinational corporations.

Worldwide Perspectives on Corporate Environmental Responsibility

The majority of international CSR studies focus on business practices and its aspects, such as business economics and the legality of environmental law. Most companies are noticing the importance of taking into account one of its most important stakeholders: employees and customers and their commitment to sustainability. Studies have demonstrated that once companies place sustainability practices they can be directly linked to financial success and customer satisfaction, which in turn can be used as a marketing tool. Although every country has a different culture, and each country determines their own scale of environmental responsibility, research has shown that there is a standard global human values that drive customer needs and wants. Companies have taken initiatives to take sustainability and align it with each company's economic goals. Managers and other people at the top, play the key role in decision-making and implementing the firm's sustainability practices.

Benefits of Corporate Environmental Responsibility

Corporate social responsibility can prove to be more profitable for companies and to extend it survivability in markets because greater awareness on this topic, in both social and business markets, has been in higher demand. Customers have responded with overall satisfaction and loyalty when companies have a better CSR, especially in countries like Spain and Brazil. Culture has an impact on the CSR ratings and studies, as well as human values across different nations.

This topic can also be found under sustainable development. This area is concerned with not only protecting the environment but maintaining economical growth. There were several agreements internationally to help adopt new business practices that held these standards, but they were considered individual and there was no law-abiding body to regulate nor implement them.

One of the other factors that is considered an integral part of sustainable development are human beings, and specific groups and their habitat. Counties and companies that more developed would lead, and other small countries and business would slowly make gains. It is important to recognize that just because corporate environmental responsibility is being recognized that consumption is something that is not discouraged.

The idea of corporate environmental responsibility (CER) is for humans to be more aware of the environmental impact and counteract their pollution/carbon footprint on the natural resources. One of the main factors is to reduce carbon footprint and carbon emissions. Many of the studies focus on trying to find a balance between economic growth and reducing waste and cleaner environments.

The environmental aspects of security have increasingly become a major issue being considered by states. The process of securitization has had a big impact in creating a new understanding of security. Globalization also plays a key role in the adoption of new environmental strategies as a multi-faceted process influencing modern societies, and creating interconnected and multidimensional environments.

CER is used by multinational corporations as well as small, local organizations. It is highlighted and more institutionalized because of stakeholders' awareness of the huge impacts of business activities on the environment. To understand CER, its relations with CSR strategies need to be recognized. CER and CSR are the main strategies that help in the creation of efficient and environmentally sustainable businesses.

Sustainability and Environmental Management

At the global scale sustainability and environmental management involves managing the oceans, freshwater systems, land and atmosphere, according to sustainability principles.

Land use change is fundamental to the operations of the biosphere because alterations in the relative proportions of land dedicated to urbanisation, agriculture, forest, woodland, grassland and pasture have a marked effect on the global water, carbon and nitrogen biogeochemical cycles. Management of the Earth's atmosphere involves assessment of all aspects of the carbon cycle to identify opportunities to address human-induced climate change and this has become a major focus of scientific research because of the potential catastrophic effects on biodiversity and human communities. Ocean circulation patterns have a strong influence on climate and weather and, in turn, the food supply of both humans and other organisms.

Atmosphere

In March 2009 at a meeting of the Copenhagen Climate Council 2,500 climate experts from 80 countries issued a keynote statement that there is now "no excuse" for failing to act on global warming and that without strong carbon reduction targets "abrupt or irreversible" shifts in climate may occur that "will be very difficult for contemporary societies to cope with". Management of the global atmosphere now involves assessment of all aspects of the carbon cycle to identify opportunities to address human-induced climate change and this has become a major focus of scientific research because of the potential catastrophic effects on biodiversity and human communities.

Other human impacts on the atmosphere include the air pollution in cities, the pollutants including toxic chemicals like nitrogen oxides, sulphur oxides, volatile organic compounds and airborne particulate matter that produce photochemical smog and acid rain, and the chlorofluorocarbons that degrade the ozone layer. Anthropogenic particulates such as sulfate aerosols in the atmosphere reduce the direct irradiance and reflectance (albedo) of the Earth's surface. Known as global dimming the decrease is estimated to have been about 4% between 1960 and 1990 although the trend has subsequently reversed. Global dimming may have disturbed the global water cycle by reducing evaporation and rainfall in some areas. It also creates a cooling effect and this may have partially masked the effect of greenhouse gases on global warming.

Oceans

Ocean circulation patterns have a strong influence on climate and weather and, in turn, the food supply of both humans and other organisms. Scientists have warned of the possibility, under the influence of climate change, of a sudden alteration in circulation patterns of ocean currents that could drastically alter the climate in some regions of the globe. Major human environmental impacts occur in the more habitable regions of the ocean fringes – the estuaries, coastline and bays. Ten per cent of the world's population – about 600 million people – live in low-lying areas vulnerable to sea level rise. Trends of concern that require management include: over-fishing (beyond sustainable levels); coral bleaching due to ocean warming and ocean acidification due to increasing levels of dissolved carbon dioxide; and sea level rise due to climate change. Because of their vastness oceans also act as a convenient dumping ground for human waste. Remedial strategies include: more careful waste management, statutory control of overfishing by adoption of sustainable fishing practices and the use of environmentally sensitive and sustainable aquaculture and fish farming, reduction of fossil fuel emissions and restoration of coastal and other marine habitat.

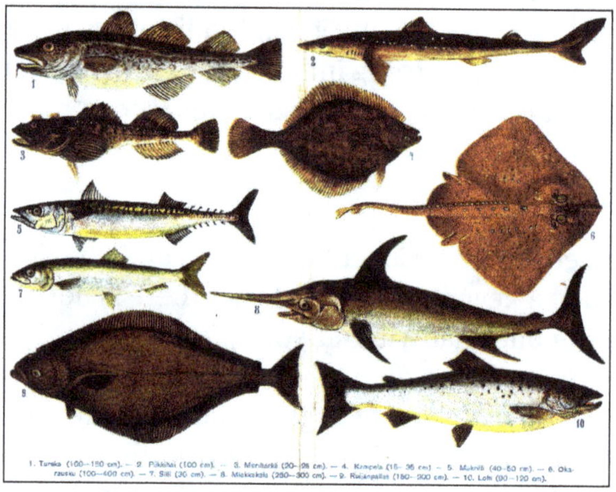

A selection of the world's saltwater fish.

Freshwater

Water covers 71% of the Earth's surface. Of this, 97.5% is the salty water of the oceans and only 2.5% freshwater, most of which is locked up in the Antarctic ice sheet. The remaining freshwater is found in lakes, rivers, wetlands, the soil, aquifers and atmosphere. All life depends on the solar-powered global water cycle, the evaporation from oceans and land to form water vapour that later condenses from clouds as rain, which then becomes the renewable part of the freshwater supply. Awareness of the global importance of preserving water for ecosystem services has only recently emerged as, during the 20th century, more than half the world's wetlands have been lost along with their valuable environmental services. Biodiversity-rich freshwater ecosystems are currently declining faster than marine or land ecosystems making them the world's most vulnerable habitats. Increasing urbanization pollutes clean water supplies and much of the world still does not have access to clean, safe water. In the industrial world demand management has slowed absolute usage rates but increasingly water is being transported over vast distances from water-rich natural areas to population-dense urban areas and energy-hungry desalination is becoming more widely used. Greater emphasis is now being placed on the improved management of blue (harvestable) and green (soil water available for plant use) water, and this applies at all scales of water management.

Land

Loss of biodiversity stems largely from the habitat loss and fragmentation produced by the human appropriation of land for development, forestry and agriculture as natural capital is progressively converted to man-made capital. Land use change is fundamental to the operations of the biosphere because alterations in the relative proportions of land dedicated to urbanisation, agriculture, forest, woodland, grassland and pasture have a marked effect on the global water, carbon and nitrogen biogeochemical cycles and this can impact negatively on both natural and human systems. At the local human scale major sustainability benefits accrue from the pursuit of green cities and sustainable parks and gardens.

Forests

Beech forest.

Since the Neolithic Revolution, human use has reduced the world's forest cover by about 47%. Present-day forests occupy about a quarter of the world's ice-free land with about half of these occurring in the tropics In temperate and boreal regions forest area is gradually increasing (with the exception of Siberia), but deforestation in the tropics is of major concern.

Forests moderate the local climate and the global water cycle through their light reflectance (albedo) and evapotranspiration. They also conserve biodiversity, protect water quality, preserve soil and soil quality, provide fuel and pharmaceuticals, and purify the air. These free ecosystem services are not given a market value under most current economic systems, and so forest conservation has little appeal when compared with the economic benefits of logging and clearance which, through soil degradation and organic decomposition returns carbon dioxide to the atmosphere. The United Nations

Food and Agriculture Organization (FAO) estimates that about 90% of the carbon stored in land vegetation is locked up in trees and that they sequester about 50% more carbon than is present in the atmosphere. Changes in land use currently contribute about 20% of total global carbon emissions (heavily logged Indonesia and Brazil are a major source of emissions). Climate change can be mitigated by sequestering carbon in reafforestation schemes, plantations and timber products. Also wood biomass can be utilized as a renewable carbon-neutral fuel. The FAO has suggested that, over the period 2005–2050, effective use of tree planting could absorb about 10–20% of man-made emissions – so monitoring the condition of the world's forests must be part of a global strategy to mitigate emissions and protect ecosystem services. However, climate change may pre-empt this FAO scenario as a study by the International Union of Forest Research Organizations in 2009 concluded that the stress of a 2.5 °C (4.5 °F) temperature rise above pre-industrial levels could result in the release of vast amounts of carbon so the potential of forests to act as carbon "sinks" is "at risk of being lost entirely".

Cultivated Land

A rice paddy. Rice, wheat, corn and potatoes make up more than half the world's food supply.

Feeding more than six billion human bodies takes a heavy toll on the Earth's resources. This begins with the appropriation of about 38% of the Earth's land surface and about 20% of its net primary productivity. Added to this are the resource-hungry activities of industrial agribusiness – everything from the crop need for irrigation water, synthetic fertilizers and pesticides to the resource costs of food packaging, transport (now a major part of global trade) and retail. Food is essential to life. But the list of environmental costs of food production is a long one: topsoil depletion, erosion and conversion to desert from constant tillage of annual crops; overgrazing; salinization; sodification; waterlogging; high levels of fossil fuel use; reliance on inorganic fertilisers and synthetic organic pesticides; reductions in genetic diversity by the mass use of monocultures; water resource depletion; pollution of waterbodies by run-off and groundwater contamination; social problems including the decline of family farms and weakening of rural communities.

All of these environmental problems associated with industrial agriculture and agribusiness are now being addressed through such movements as sustainable agriculture, organic farming and more sustainable business practices.

Extinctions

The extinct dodo (Raphus cucullatus).

Although biodiversity loss can be monitored simply as loss of species, effective conservation demands the protection of species within their natural habitats and ecosystems. Following human migration and population growth, species extinctions have progressively increased to a rate unprecedented since the Cretaceous–Paleogene extinction event. Known as the Holocene extinction event this current human-induced extinction of species ranks as one of the world's six mass extinction events. Some scientific estimates indicate that up to half of presently existing species may become extinct by 2100. Current extinction rates are 100 to 1000 times their prehuman levels with more than 10% birds and mammals threatened, about 8% of plants, 5% of fish and more than 20% of freshwater species.

The 2008 IUCN Red List warns that long-term droughts and extreme weather put additional stress on key habitats and, for example, lists 1,226 bird species as threatened with extinction, which is one-in-eight of all bird species. The Red List Index also identifies 44 tree species in Central Asia as under threat of extinction due to over-exploitation and human development and threatening the region's forests which are home to more than 300 wild ancestors of modern domesticated fruit and nut cultivars.

Biological Invasions

Kudzu (*Pueraria lobata*) infesting trees.

In many parts of the industrial world land clearing for agriculture has diminished and here the greatest threat to biodiversity, after climate change, has become the destructive effect of invasive species. Increasingly efficient global transport has facilitated the spread of organisms across the planet. The potential danger of this aspect of globalization is starkly illustrated through the spread of human diseases like HIV AIDS, mad cow disease, bird flu and swine flu, but invasive plants and animals are also having a devastating impact on native biodiversity. Non-indigenous organisms can quickly occupy disturbed land and natural areas where, in the absence of their natural predators, they are able to thrive. At the global scale this issue is being addressed through the Global Invasive Species Information Network but there is improved international biosecurity legislation to minimise the transmission of pathogens and invasive organisms. Also, through CITES legislation there is control the trade in rare and threatened species. Increasingly at the local level public awareness programs are alerting communities, gardeners, the nursery industry, collectors, and the pet and aquarium industries, to the harmful effects of potentially invasive species.

Resistance to Change

The environmental sustainability problem has proven difficult to solve. The modern environmental movement has attempted to solve the problem in a large variety of ways. But little progress has been made, as shown by severe ecological footprint overshoot and lack of sufficient progress on the climate change problem. Something within the human system in preventing change to a sustainable mode of behavior. That system trait is systemic change resistance. Change resistance is also known as organizational resistance, barriers to change, or policy resistance.

Ecosystem Services

Honey bee on Avocado crop. Pollination is just one type of ecosystem service.

Upland bog in Wales, forming the official source of the River Severn. Healthy bogs sequester carbon, hold back water thereby reducing flood risk, and supply cleaned water better than degraded habitats do.

Social forestry in Andhra Pradesh, India, providing fuel, soil protection, shade and even well-being to travellers.

Ecosystem services are the many and varied benefits that humans freely gain from the natural environment and from properly-functioning ecosystems. Such ecosystems include, for example, agroecosystems, forest ecosystems, grassland ecosystems and aquatic ecosystems. These ecosystems functioning properly provides such things like agricultural produce, timber, and aquatic organisms such as fishes and crabs. Collectively, these benefits are becoming known as 'ecosystem services', and are often integral to the provisioning of clean drinking water, the decomposition of wastes, and the natural pollination of crops and other plants.

While scientists and environmentalists have discussed ecosystem services implicitly for decades, the Millennium Ecosystem Assessment (MA) in the early 2000s popularized this concept. There, ecosystem services are grouped into four broad categories: *provisioning*, such as the production of food and water; *regulating*, such as the control of climate and disease; *supporting*, such as nutrient cycles and oxygen production; and *cultural*, such as spiritual and recreational benefits. To help inform decision-makers, many ecosystem services are being assigned economic values.

While the notion of human dependence on Earth's ecosystems reaches to the start of *Homo sapiens*' existence, the term 'natural capital' was first coined by E.F. Schumacher in 1973 in his book Small is Beautiful. Recognition of how ecosystems could provide complex services to humankind date back to at least Plato (c. 400 BC) who understood that deforestation could lead to soil erosion and the drying of springs. Modern ideas of ecosystem services probably began when Marsh challenged in 1864 the idea that Earth's natural resources are unbounded by pointing out changes in soil fertility in the Mediterranean. It was not until the late 1940s that three key authors—Henry Fairfield Osborn, Jr, William Vogt, and Aldo Leopold —promoted recognition of human dependence on the environment.

In 1956, Paul Sears drew attention to the critical role of the ecosystem in processing wastes and recycling nutrients. Paul Ehrlich and Rosa Weigert called attention to "ecological systems" in their environmental science textbook and "the most subtle and dangerous threat to man's existence the potential destruction, by man's own activities, of those ecological systems upon which the very existence of the human species depends".

The term "environmental services" was introduced in a 1970 report of the Study of Critical Environmental Problems, which listed services including insect pollination, fisheries, climate regulation and flood control. In following years, variations of the term were used, but eventually 'ecosystem services' became the standard in scientific literature.

The ecosystem services concept has continued to expand and includes socio-economic and conservation objectives. A history of the concepts and terminology of ecosystem services as of 1997, can be found in Daily's book "Nature's Services: Societal Dependence on Natural Ecosystems".

While Gretchen Daily's original definition distinguished between ecosystem goods and ecosystem services, Robert Costanza and colleagues' later work and that of the Millennium Ecosystem Assessment lumped all of these together as ecosystem services.

Per the 2006 Millennium Ecosystem Assessment (MA), ecosystem services are "the benefits people obtain from ecosystems". The MA also delineated the four categories of ecosystem services—supporting, provisioning, regulating and cultural.

By 2010, there had evolved various working definitions and descriptions of ecosystem services in the literature. To prevent double counting in ecosystem services audits, for instance, The Economics of Ecosystems and Biodiversity (TEEB) replaced "Supporting Services" in the MA with "Habitat Services" and "ecosystem functions", defined as "a subset of the interactions between ecosystem structure and processes that underpin the capacity of an ecosystem to provide goods and services".

Categorization

The Millennium Ecosystem Assessment (MA) report 2005 defines *Ecosystem services* as benefits people obtain from ecosystems and distinguishes four categories of ecosystem services, where the so-called supporting services are regarded as the basis for the services of the other three categories.

Detritivores like this dung beetle help to turn animal wastes into organic material that can be reused by primary producers.

Supporting Services

These include services such as nutrient cycling, primary production, soil formation, habitat provision and pollination. These services make it possible for the ecosystems to continue providing services such as food supply, flood regulation, and water purification.

Provisioning Services

- Food (including seafood and game), crops, wild foods, and spices.
- Raw materials (including lumber, skins, fuel wood, organic matter, fodder, and fertilizer).
- Genetic resources (including crop improvement genes, and health care).
- Water.
- Biogenic minerals.
- Medicinal resources (including pharmaceuticals, chemical models, and test and assay organisms).
- Energy (hydropower, biomass fuels).
- Ornamental resources (including fashion, handicraft, jewelry, pets, worship, decoration and souvenirs like furs, feathers, ivory, orchids, butterflies, aquarium fish, shells, etc).

Regulating Services

- Carbon sequestration and climate regulation.
- Predation regulates prey populations.

- Waste decomposition and detoxification.
- Purification of water and air.
- Pest and disease control.

Cultural Services

- Cultural (including use of nature as motif in books, film, painting, folklore, national symbols, architect, advertising, etc).
- Spiritual and historical (including use of nature for religious or heritage value or natural).
- Recreational experiences (including ecotourism, outdoor sports, and recreation).
- Science and education (including use of natural systems for school excursions, and scientific discovery).
- Therapeutic (including Ecotherapy, social forestry and animal assisted therapy).

The concept of cultural ecosystem services can be operationalized. A good review of approaches in landscape aesthetics, cultural heritage, outdoor recreation, and spiritual significance to define and assess cultural values of our environment so that they fit into the ecosystem services approach is given by Daniel et al. who vote for models that explicitly link ecological structures and functions with cultural values and benefits. There also is a fundamental *critique* of the concept of cultural ecosystem services that builds on three arguments:

- Pivotal cultural values attaching to the natural/cultivated environment rely on an area's unique character that cannot be addressed by methods that use universal scientific parameters to determine ecological structures and functions.
- If a natural/cultivated environment has symbolic meanings and cultural values the object of these values are not ecosystems but shaped phenomena like mountains, lakes, forests, and, mainly, symbolic landscapes.
- Those cultural values do result not from properties produced by ecosystems but are the product of a specific way of seeing within the given cultural framework of symbolic experience.

Common International Classification of Ecosystem Services

The Common International Classification of Ecosystem Services (CICES) is a classification scheme developed to acounting systems (like National counts etc.), in order to avoid double-counting of Suporting Services with others Provisioning and Regulating Services.

Examples:

The following examples illustrate the relationships between humans and natural ecosystems through the services derived from them:

- In New York City, where the quality of drinking water had fallen below standards required by the U.S. Environmental Protection Agency (EPA), authorities opted to restore the polluted Catskill Watershed that had previously provided the city with the ecosystem service of water purification. Once the input of sewage and pesticides to the watershed area was reduced, natural abiotic processes such as soil absorption and filtration of chemicals, together with biotic recycling via root systems and soil microorganisms, water quality improved to levels that met government standards. The cost of this investment in natural capital was estimated between $1–1.5 billion, which contrasted dramatically with the estimated $6–8 billion cost of constructing a water filtration plant plus the $300 million annual running costs.
- Pollination of crops by bees is required for 15–30% of U.S. food production; most large-scale farmers import non-native honey bees to provide this service. One study reports that in California's agricultural region, it was found that wild bees alone could provide partial or complete pollination services or enhance the services provided by honey bees through behavioral interactions. However, intensified agricultural practices can quickly erode pollination services through the loss of species. The remaining species are unable to compensate this. The results of this study also indicate that the proportion of chaparral and oak-woodland habitat available for wild bees within 1–2 km of a farm can stabilize and enhance the provision of pollination services. The presence of such ecosystem elements functions almost like an insurance policy for farmers.
- In watersheds of the Yangtze River (China), spatial models for water flow through different forest habitats were created to determine potential contributions for hydroelectric power in the region. By quantifying the relative value of ecological parameters (vegetation-soil-slope complexes), researchers were able to estimate the annual economic benefit of maintaining forests in the watershed for power services to be 2.2 times that if it were harvested once for timber.
- In the 1980s, mineral water company Vittel (now a brand of Nestlé Waters) faced a critical problem. Nitrates and pesticides were entering the company's springs in northeastern France. Local farmers had intensified agricultural practices and cleared native vegetation that previously had filtered water before it seeped into the aquifer used by Vittel. This contamination threatened the company's right to use the "natural mineral water" label under French law. In response to this business risk, Vittel developed an incentive package for farmers to improve their agricultural practices and consequently reduce water pollution

that had affected Vittel's product. For example, Vittel provided subsidies and free technical assistance to farmers in exchange for farmers' agreement to enhance pasture management, reforest catchments, and reduce the use of agrochemicals. This is an example of a payment for ecosystem services program.

- It was counted that to plant 15 000 ha new woodland in the UK, if we consider only the value of timber, it would cost £79 000 000 which is more than the benefit of £65 000 000. If, however, we include all other benefits the trees in lowland could provide (like soil stabilization, wind deflection, recreation, food production, air purification, carbon storage, wildlife habitat, fuel production, cooling, flood prevention), the costs will increase due to displacing the profitable farmland (would be around £231 000 000) but will be overweight by benefits of £546 000 000.
- In Europe, various projects are implemented in order to define the values of concrete ecosystems and to implement this concept into decision making process. For example, "LIFE Viva grass" project aims to do this with grasslands in Baltics.

Ecology

Understanding of ecosystem services requires a strong foundation in ecology, which describes the underlying principles and interactions of organisms and the environment. Since the scales at which these entities interact can vary from microbes to landscapes, milliseconds to millions of years, one of the greatest remaining challenges is the descriptive characterization of energy and material flow between them. For example, the area of a forest floor, the detritus upon it, the microorganisms in the soil and characteristics of the soil itself will all contribute to the abilities of that forest for providing ecosystem services like carbon sequestration, water purification, and erosion prevention to other areas within the watershed. Note that it is often possible for multiple services to be bundled together and when benefits of targeted objectives are secured, there may also be ancillary benefits—the same forest may provide habitat for other organisms as well as human recreation, which are also ecosystem services.

The complexity of Earth's ecosystems poses a challenge for scientists as they try to understand how relationships are interwoven among organisms, processes and their surroundings. As it relates to human ecology, a suggested research agenda for the study of ecosystem services includes the following steps:

- Identification of ecosystem service providers (ESPs)—species or populations that provide specific ecosystem services—and characterization of their functional roles and relationships.
- Determination of community structure aspects that influence how ESPs function in their natural landscape, such as compensatory responses that stabilize function and non-random extinction sequences which can erode it.

- Assessment of key environmental (abiotic) factors influencing the provision of services.
- Measurement of the spatial and temporal scales ESPs and their services operate on.

Recently, a technique has been developed to improve and standardize the evaluation of ESP functionality by quantifying the relative importance of different species in terms of their efficiency and abundance. Such parameters provide indications of how species respond to changes in the environment (i.e. predators, resource availability, climate) and are useful for identifying species that are disproportionately important at providing ecosystem services. However, a critical drawback is that the technique does not account for the effects of interactions, which are often both complex and fundamental in maintaining an ecosystem and can involve species that are not readily detected as a priority. Even so, estimating the functional structure of an ecosystem and combining it with information about individual species traits can help us understand the resilience of an ecosystem amidst environmental change.

Many ecologists also believe that the provision of ecosystem services can be stabilized with biodiversity. Increasing biodiversity also benefits the variety of ecosystem services available to society. Understanding the relationship between biodiversity and an ecosystem's stability is essential to the management of natural resources and their services.

Redundancy Hypothesis

The concept of ecological redundancy is sometimes referred to as *functional compensation* and assumes that more than one species performs a given role within an ecosystem. More specifically, it is characterized by a particular species increasing its efficiency at providing a service when conditions are stressed in order to maintain aggregate stability in the ecosystem. However, such increased dependence on a compensating species places additional stress on the ecosystem and often enhances its susceptibility to subsequent disturbance. The redundancy hypothesis can be summarized as "species redundancy enhances ecosystem resilience".

Another idea uses the analogy of rivets in an airplane wing to compare the exponential effect the loss of each species will have on the function of an ecosystem; this is sometimes referred to as *rivet popping*. If only one species disappears, the loss of the ecosystem's efficiency as a whole is relatively small; however, if several species are lost, the system essentially collapses–similar to an airplane that lost too many rivets. The hypothesis assumes that species are relatively specialized in their roles and that their ability to compensate for one another is less than in the redundancy hypothesis. As a result, the loss of any species is critical to the performance of the ecosystem. The key difference is the rate at which the loss of species affects total ecosystem functioning.

Portfolio Effect

A third explanation, known as the portfolio effect, compares biodiversity to stock holdings, where diversification minimizes the volatility of the investment, or in this case, the risk of instability of ecosystem services. This is related to the idea of response diversity where a suite of species will exhibit differential responses to a given environmental perturbation. When considered together, they create a stabilizing function that preserves the integrity of a service.

Several experiments have tested these hypotheses in both the field and the lab. In ECOTRON, a laboratory in the UK where many of the biotic and abiotic factors of nature can be simulated, studies have focused on the effects of earthworms and symbiotic bacteria on plant roots. These laboratory experiments seem to favor the rivet hypothesis. However, a study on grasslands at Cedar Creek Reserve in Minnesota supports the redundancy hypothesis, as have many other field studies.

Economics

Sustainable urban drainage pond near housing in Scotland. The filtering and cleaning of surface and waste water by natural vegetation is a form of ecosystem service.

There are questions regarding the environmental and economic values of ecosystem services. Some people may be unaware of the environment in general and humanity's interrelatedness with the natural environment, which may cause misconceptions. Although environmental awareness is rapidly improving in our contemporary world, ecosystem capital and its flow are still poorly understood, threats continue to impose, and we suffer from the so-called 'tragedy of the commons'. Many efforts to inform decision-makers of current versus future costs and benefits now involve organizing and translating scientific knowledge to economics, which articulate the consequences of our choices in comparable units of impact on human well-being. An especially challenging aspect of this process is that interpreting ecological information collected from one spatial-temporal scale does not necessarily mean it can be applied at another; understanding the dynamics of ecological processes relative to ecosystem services is essential in aiding economic decisions. Weighting factors such as a service's irreplaceability or bundled services can also allocate economic value such that goal attainment becomes more efficient.

The economic valuation of ecosystem services also involves social communication and information, areas that remain particularly challenging and are the focus of many researchers. In general, the idea is that although individuals make decisions for any variety of reasons, trends reveal the aggregative preferences of a society, from which the economic value of services can be inferred and assigned. The six major methods for valuing ecosystem services in monetary terms are:

- Avoided cost: Services allow society to avoid costs that would have been incurred in the absence of those services (e.g. waste treatment by wetland habitats avoids health costs).
- Replacement cost: Services could be replaced with man-made systems (e.g. restoration of the Catskill Watershed cost less than the construction of a water purification plant).
- Factor income: Services provide for the enhancement of incomes (e.g. improved water quality increases the commercial take of a fishery and improves the income of fishers).
- Travel cost: Service demand may require travel, whose costs can reflect the implied value of the service (e.g. value of ecotourism experience is at least what a visitor is willing to pay to get there).
- Hedonic pricing: Service demand may be reflected in the prices people will pay for associated goods (e.g. coastal housing prices exceed that of inland homes).
- Contingent valuation: Service demand may be elicited by posing hypothetical scenarios that involve some valuation of alternatives (e.g. visitors willing to pay for increased access to national parks).

Estuarine and Coastal Ecosystem Services

Ecosystem services are defined as the gains acquired by humankind from surroundings ecosystems. Four different types of ecosystem services have been distinguished by the scientific body: regulating services, provisioning services, cultural services and supporting services. An ecosystem does not necessarily offer all four types of services simultaneously; but given the intricate nature of any ecosystem, it is usually assumed that humans benefit from a combination of these services. The services offered by diverse types of ecosystems (forests, seas, coral reefs, mangroves, etc.) differ in nature and in consequence. In fact, some services directly affect the livelihood of neighboring human populations (such as fresh water, food or aesthetic value, etc.) while other services affect general environmental conditions by which humans are indirectly impacted (such as climate change, erosion regulation or natural hazard regulation, etc.).

Estuarine and coastal ecosystems are both marine ecosystems. An estuary is defined as the area in which a river meets the sea or the ocean. The waters surrounding this area are predominantly salty waters or brackish waters; and the incoming river water is dynamically motioned by the tide. An estuary strip may be covered by populations of reed (or similar plants) and/or sandbanks (or similar form or land).

A coastal ecosystem occurs in areas where the sea or ocean waters meet the land.

Regulating Services

Regulating services are the "benefits obtained from the regulation of ecosystem processes". In the case of coastal and estuarine ecosystems, these services include climate regulation, waste treatment and disease control and natural hazard regulation.

Climate Regulation

Both the biotic and abiotic ensembles of marine ecosystems play a role in climate regulation. They act as sponges when it comes to gases in the atmosphere, retaining large levels of CO_2 and other greenhouse gases (methane and nitrous oxide). Marine plants also use CO_2 for photosynthesis purposes and help in reducing the atmospheric CO_2. The oceans and seas absorb the heat from the atmosphere and redistribute it through the means of water currents, and atmospheric processes, such as evaporation and the reflection of light allow for the cooling and warming of the overlying atmosphere. The ocean temperatures are thus imperative to the regulation of the atmospheric temperatures in any part of the world: "without the ocean, the Earth would be unbearably hot during the daylight hours and frigidly cold, if not frozen, at night".

Waste Treatment and Disease Regulation

Another service offered by marine ecosystem is the treatment of wastes, thus helping in the regulation of diseases. Wastes can be diluted and detoxified through transport across marine ecosystems; pollutants are removed from the environment and stored, buried or recycled in marine ecosystems: "Marine ecosystems break down organic waste through microbial communities that filter water, reduce/limit the effects of eutrophication, and break down toxic hydrocarbons into their basic components such as carbon dioxide, nitrogen, phosphorus, and water". The fact that waste is diluted with large volumes of water and moves with water currents leads to the regulation of diseases and the reduction of toxics in seafood.

Buffer Zones

Coastal and estuarine ecosystems act as buffer zones against natural hazards and environmental disturbances, such as floods, cyclones, tidal surges and storms. The role they play is to "[absorb] a portion of the impact and thus [lessen] its effect on the land". Wetlands, for example, and the vegetation it supports – trees, root mats, etc. – retain large amounts of water (surface water, snowmelt, rain, groundwater) and then slowly releases them back, decreasing the likeliness of floods. Mangrove forests protect coastal shorelines from tidal erosion or erosion by currents; a process that was studied after the 1999 cyclone that hit India. Villages that were surrounded with mangrove forests encountered less damages than other villages that weren't protected by mangroves.

Provisioning Services

Provisioning services consist of all "the products obtained from ecosystems". Marine ecosystems provide people with: wild & cultured seafood, fresh water, fiber & fuel and biochemical & genetic resources.

Marine Products

Humans consume a large number of products originating from the seas, whether as a nutritious product or for use in other sectors: "More than one billion people worldwide, or one-sixth of the global population, rely on fish as their main source of animal protein. In 2000, marine and coastal fisheries accounted for 12 percent of world food production". Fish and other edible marine products – primarily fish, shellfish, roe and seaweeds – constitute for populations living along the coast the main elements of the local cultural diets, norms and traditions. A very pertinent example would be sushi, the national food of Japan, which consists mostly of different types of fish and seaweed.

Fresh Water

Water bodies that are not highly concentrated in salts are referred to as 'fresh water' bodies. Fresh water may run through lakes, rivers and streams, to name a few; but it is most prominently found in the frozen state or as soil moisture or buried deep underground. Fresh water is not only important for the survival of humans, but also for the survival of all the existing species of animals, plants.

Raw Materials

Marine creatures provide us with the raw materials needed for the manufacturing of clothing, building materials (lime extracted from coral reefs), ornamental items and personal-use items (luffas, art and jewelry): "The skin of marine mammals for clothing, gas deposits for energy production, lime (extracted from coral reefs) for building construction, and the timber of mangroves and coastal forests for shelter are some of the more familiar uses of marine organisms. Raw marine materials are utilized for non-essential goods as well, such as shells and corals in ornamental items". Humans have also referred to processes within marine environments for the production of renewable energy: using the power of waves – or tidal power – as a source of energy for the powering of a turbine, for example. Oceans and seas are used as sites for offshore oil and gas installations, offshore wind farms.

Biochemical and Genetic Resources

Biochemical resources are compounds extracted from marine organisms for use in medicines, pharmaceuticals, cosmetics and other biochemical products. Genetic resources are the genetic information found in marine organisms that would later on be used for animal and plant breeding and for technological advances in the biological field. These resources are either directly taken out from an organism – such as fish oil as a source of omega3 –, or used as a model for innovative man-made products: "such as the construction of fiber optics technology based on the properties of sponges. Compared to terrestrial products, marine-sourced products tend to be more highly bioactive, likely due to the fact that marine organisms have to retain their potency despite being diluted in the surrounding sea-water".

Cultural Services

Cultural services relate to the non-material world, as they benefit the benefit recreational, aesthetic, cognitive and spiritual activities, which are not easily quantifiable in monetary terms.

Inspirational

Marine environments have been used by many as an inspiration for their works of art, music, architecture, traditions Water environments are spiritually important as a lot of people view them as a means for rejuvenation and change of perspective. Many also consider the water as being a part of their personality, especially if they have lived near it since they were kids: they associate it to fond memories and past experiences. Living near water bodies for a long time results in a certain set of water activities that become a ritual in the lives of people and of the culture in the region.

Recreation and Tourism

Sea sports are very popular among coastal populations: surfing, snorkeling, whale watching, kayaking, recreational fishing a lot of tourists also travel to resorts close to the sea or rivers or lakes to be able to experience these activities, and relax near the water.

Beach accommodated into a recreational area.

Science and Education

A lot can be learned from marine processes, environments and organisms – that could be implemented into our daily actions and into the scientific domain. Although much is still yet to still be known about the ocean world: "by the extraordinary intricacy and complexity of the marine environment and how it is influenced by large spatial scales, time lags, and cumulative effects".

Supporting Services

Supporting services are the services that allow for the other ecosystem services to be present. They have indirect impacts on humans that last over a long period of time. Several services can be considered as being both supporting services and regulating/cultural/provisioning services.

Nutrient Cycling

Nutrient cycling is the movement of nutrients through an ecosystem by biotic and abiotic processes. The ocean is a vast storage pool for these nutrients, such as carbon, nitrogen and phosphorus. The nutrients are absorbed by the basic organisms of the marine food web and are thus transferred from one organism to the other and from one ecosystem to the other. Nutrients are recycled through the life cycle of organisms as they die and decompose, releasing the nutrients into the neighboring environment. "The service of nutrient cycling eventually impacts all other ecosystem services as all living things require a constant supply of nutrients to survive".

Biologically Mediated Habitats

Biologically mediated habitats are defined as being the habitats that living marine structures offer to other organisms. These need not to be designed for the sole purpose of serving as a habitat, but happen to become living quarters whilst growing naturally. For example, coral reefs and mangrove forests are home to numerous species of fish, seaweed and shellfish. The importance of these habitats is that they allow for interactions between different species, aiding the provisioning of marine goods and services. They are also very important for the growth at the early life stages of marine species (breeding and bursary spaces), as they serve as a food source and as a shelter from predators.

Coral and other living organisms serve as habitats for many marine species.

Primary Production

Primary production refers to the production of organic matter, i.e., chemically bound energy, through processes such as photosynthesis and chemosynthesis. The organic matter produced by primary producers forms the basis of all food webs. Further, it generates oxygen (O_2), a molecule necessary to sustain animals and humans. On average, a human consumes about 550 liter of oxygen per day, whereas plants produce 1,5 liter of oxygen per 10 grams of growth.

Land use Change Decisions

Ecosystem services decisions require making complex choices at the intersection of ecology, technology, society and the economy. The process of making ecosystem services decisions must consider the interaction of many types of information, honor all stakeholder viewpoints, including regulatory agencies, proposal proponents, decision makers, residents, NGOs, and measure the impacts on all four parts of the intersection. These decisions are usually spatial, always multi-objective, and based on uncertain data, models, and estimates. Often it is the combination of the best science combined with the stakeholder values, estimates and opinions that drive the process.

One analytical study modeled the stakeholders as agents to support water resource management decisions in the Middle Rio Grande basin of New Mexico. This study focused on modeling the stakeholder inputs across a spatial decision, but ignored uncertainty. Another study used Monte Carlo methods to exercise econometric models of landowner decisions in a study of the effects of land-use change. Here the stakeholder inputs were modeled as random effects to reflect the uncertainty. A third study used a Bayesian decision support system to both model the uncertainty in the scientific information Bayes Nets and to assist collecting and fusing the input from stakeholders. This study was about siting wave energy devices off the Oregon Coast, but presents a general method for managing uncertain spatial science and stakeholder information in a decision making environment. Remote sensing data and analyses can be used to assess the health and extent of land cover classes that provide ecosystem services, which aids in planning, management, monitoring of stakeholders' actions, and communication between stakeholders.

In Baltic countries scientists, nature conservationists and local authorities are implementing integrated planning approach for grassland ecosystems. They are developing Integrated Planning Tool that will be based on GIS (geographic information system) technology and put online that will help for planners to choose the best grassland management solution for concrete grassland. It will look holistically at the processes in the countryside and help to find best grassland management solutions by taking into account both natural and socioeconomic factors of the particular site.

Habitat Conservation

Habitat conservation for wild species is one of the most important issues facing the environment today – both in the ocean and on land.

Habitat conservation for wild species is one of the most important issues facing the environment today – both in the ocean and on land. As human populations increase, land use increases, and wild species have smaller spaces to call home. More than half of Earth's terrestrial surface has been altered due to human activity, resulting in drastic deforestation, erosion and loss of topsoil, biodiversity loss, and extinction. Species cannot survive outside of their natural habitat without human intervention, such as the habitats found in a zoo or aquarium, for example. Preserving habitats is essential to preserving biodiversity. Migratory species are particularly vulnerable to habitat destruction because they tend to inhabit more than one natural habitat. This creates the need to not only preserve the two habitats for migratory species, but also their migratory route. Altering a natural habitat even slightly can result in a domino effect that harms the entire ecosystem.

Problems

Habitat destruction is a huge problem in the marine environment. Habitats are destroyed by:

- Destructive fishing activity: Bottom trawling and dynamiting coral reefs destroy entire ecosystems.
- Coastal development: Habitats are destroyed when marshes are dredged for real estate development. Soil runoff and erosion result in excess nutrients from fertilizers and domestic sewage, which then leads to harmful algae blooms that block sunlight and deplete the water of oxygen. It also causes silt to build-up on coral reefs, which blocks sunlight necessary for coral to grow.
- Pollution: Development near coastal waters contaminates the Ocean with toxic substances, such as industrial chemicals, pesticides, and motor oil.

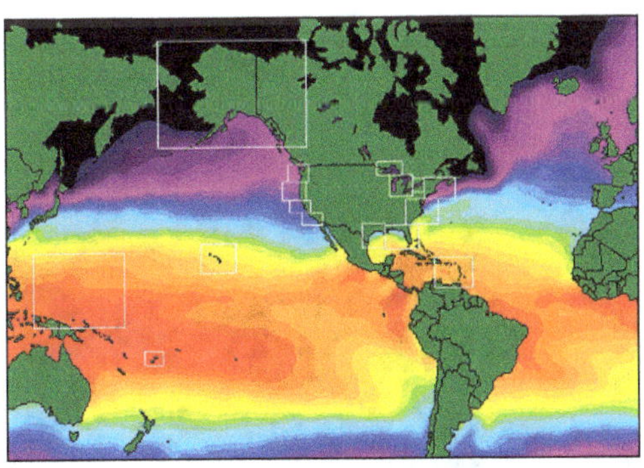

- Dredging ship channels: Removes accumulated sediment and pollutants, re-suspending them into the water. Dredging can also destroy sea grass beds and other habitats that provide food, shelter, and breeding grounds. The dredged material must be disposed of, and is often dumped into salt marshes, damaging very productive marine habitats in the process.

Solutions

- Marine Protected Areas (MPAs): Marine sites such as sanctuaries, fisheries management areas, state conservation areas, and wildlife refuges established to protect habitats, endangered species, and to restore the health of marine ecosystems in areas jeopardized by habitat and species loss. Examples: NOAA National Marine Sanctuaries: USA.
- Marine Reserves: Marine sites that provide a higher degree of ecosystem protection by prohibiting fishing, mineral extraction, and other habitat-altering activities. Marine Reserves are far more effective than MPAs, but unfortunately, they are not as common. Example: Marine Reserves in New Zealand.
- Land use and development regulation: An integrated approach to land use and management based on scientific knowledge is needed to protect coastal areas. Policy makers need to be informed on the impact coastal development is having on marine habitats through accessible and evidence-based information.
- Monitoring and reporting: Some conservation efforts are empowering the citizens with the responsibility for monitoring water quality in their coastal communities through sampling and testing, photographing fouled areas, and providing information to local policy makers for action.
- Zoning: Related to integrated land use and development management, zoning coastal areas into MPAs, Marine Reserves, approved fishing areas, with varying levels of use has the potential to slow some of the habitat degradation caused by development. The Great Barrier Reef is managed in this way. Through cooperation among local, state, and national governments, this approach may provide a viable solution to all stakeholders from tourists, to the fishing industry, to conservation efforts, etc.

Although habitat destruction has been increasing for many years, the protection of marine habitats has only recently become an issue of critical importance to conservation efforts, local and national governments, and international marine conservation groups. The Ocean's invulnerability to human activity is now being realized as a myth. Coastal regions are still experiencing intense pressure by exploding coastal populations; however, there are solutions at hand to prevent further damage from occurring.

Everyone knows that the northern spotted owl is threatened because of destruction to the forests of the Pacific Northwest, but what will happen to the Pacific seahorse if its habitat continues to decline? Due to the lack of a strong public sea ethic, marine life doesn't appear on the conservation radar screen as much as its terrestrial counterparts, but ocean habitats are in decline as well, and therefore, the creatures they support are too, which in turn support us. Most marine habitat destruction is caused by pollution, commercial fishing equipment, coastal development, and other human activity. Much of it can be avoided with simple measures.

Secondary Spill Containment

Secondary spill containment is the containment of hazardous liquids in order to prevent pollution of soil and water. Common techniques include the use of spill berms to contain oil-filled equipment, fuel tanks, truck washing decks, or any other places or items that may leak hazardous liquids.

Secondary spill containment involves the sequestration of hazardous waste to prevent the contamination of local soils and water.

Adaptable Secondary Oil Containment around a Transformer.

United States Environmental Protection Agency (EPA) Spill Prevention, Control, and Countermeasure (SPCC) guidelines require that facilities that store large quantities of petroleum (products) must have a plan in place to contain a spill. The purpose of the SPCC rule is to establish requirements for facilities to prevent a discharge of oil into navigable waters or adjoining shorelines. Within the electric utility industry, oil-filled transformers are often in need of secondary containment. Outdated secondary containment techniques such as concrete catch-basins are quickly losing ground to solutions that offer more cost-effective cleanup in case of a spill or leak. One example of a more cost-effective method involves placing a geotextile boom filled with oil solidifying polymers around a transformer. These geotextile barriers allow for flow of water, but completely solidify oil in the event of a leak and effectively seal the spill. Many electrical utility companies are switching to this method because it saves them significant amounts of money when a spill occurs, because there is no need to employ vac-trucks afterwards to clean up a spill inside a catch-basin.

Portable Spill Containment

Portable spill containment around fracking equipment.

Portable containment berms are essentially a basin that can catch many different types of hazardous liquids and chemicals. They are a form of secondary spill containment useful for containing mobile equipment such as oil drums, trucks, tankers and trailers. Unlike geotextile berms, portable berms usually do not solidify oil. Many companies involved in fracking use spill containment berms to capture contaminated water that is a by-product of the operation. Each well site has multiple trucks that transport water used in deep well drilling procedure.

Life-cycle Assessment

Life-cycle assessment (LCA, also known as life-cycle analysis, ecobalance, and cradle-to-grave analysis) is a technique to assess environmental impacts associated with all the stages of a product's life from raw material extraction through materials processing, manufacture, distribution, use, repair and maintenance, and disposal or recycling. Designers use this process to help critique their products. LCAs can help avoid a narrow outlook on environmental concerns by:

- Compiling an inventory of relevant energy and material inputs and environmental releases.
- Evaluating the potential impacts associated with identified inputs and releases.

The goal of LCA is to compare the full range of environmental effects assignable to products and services by quantifying all inputs and outputs of material flows and assessing how these material flows affect the environment. This information is used to improve processes, support policy and provide a sound basis for informed decisions.

The term life cycle refers to the notion that a fair, holistic assessment requires the assessment of raw-material production, manufacture, distribution, use and disposal including all intervening transportation steps necessary or caused by the product's existence.

There are two main types of LCA. Attributional LCAs seek to establish (or attribute) the burdens associated with the production and use of a product, or with a specific service or process, at a point in time. Consequential LCAs seek to identify the environmental consequences of a decision or a proposed change in a system under study, which means that market and economic implications of a decision may have to be taken into account. Social LCA is under development as a different approach to life cycle thinking intended to assess social implications or potential impacts. Social LCA should be considered as an approach that is complementary to environmental LCA.

The procedures of life cycle assessment (LCA) are part of the ISO 14000 environmental management standards: in ISO 14040:2006 and 14044:2006. (ISO 14044 replaced earlier versions of ISO 14041 to ISO 14043.) GHG product life cycle assessments can also comply with specifications such as PAS 2050 and the GHG Protocol Life Cycle Accounting and Reporting Standard.

Four Main Phases

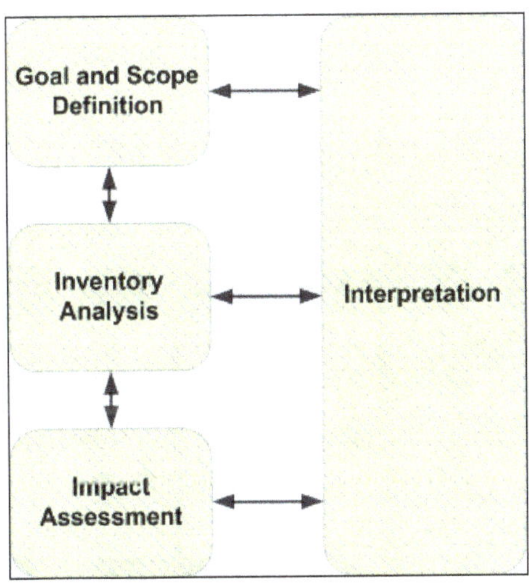

Illustration of LCA phases.

According to the ISO 14040 and 14044 standards, a Life Cycle Assessment is carried out in four distinct phases as illustrated in the figure shown to the right. The phases are often interdependent in that the results of one phase will inform how other phases are completed.

Goal and Scope

An LCA starts with an explicit statement of the goal and scope of the study, which sets out the context of the study and explains how and to whom the results are to be communicated. This is a key step and the ISO standards require that the goal and scope of an LCA be clearly defined and consistent with the intended application. The goal and scope document, therefore, includes technical details that guide subsequent work:

- The functional unit, which defines what precisely is being studied and quantifies the service delivered by the product system, providing a reference to which the inputs and outputs can be related. Further, the functional unit is an important basis that enables alternative goods, or services, to be compared and analyzed. So to explain this a functional system which is inputs, processes and outputs contains a functional unit, that fulfills a function, for example paint is covering a wall, making a functional unit of 1m² covered for 10 years. The functional flow would be the items necessary for that function, so this would be a brush, tin of paint and the paint itself.
- The system boundaries; which are delimitations of which processes that should be included in the analysis of a product system.
- Any assumptions and limitations.
- The allocation methods used to partition an environmental load of a process when several products or functions share the same process; allocation is commonly dealt with in one of three ways: system expansion, substitution, and partition. Doing this is not easy and different methods may give different results.
- The impact categories chosen for example human toxicity, smog, global warming, eutrophication.

Life-cycle Inventory

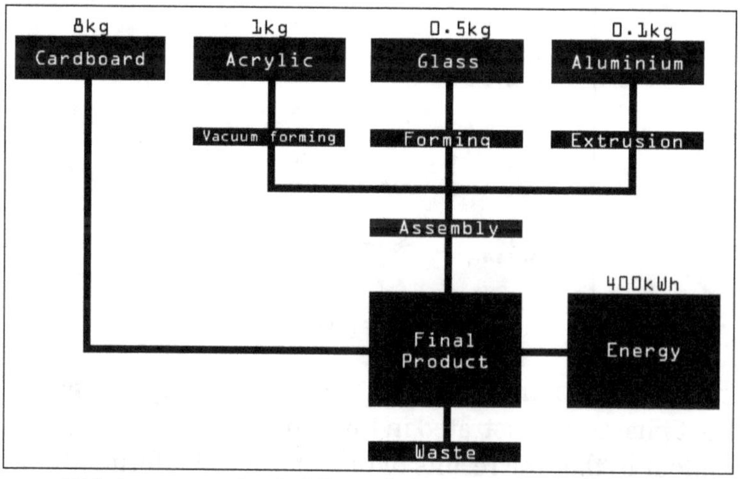

This is an example of a Life-cycle inventory (LCI) diagram.

Life Cycle Inventory (LCI) analysis involves creating an inventory of flows from and to nature for a product system. Inventory flows include inputs of water, energy, and raw materials, and releases to air, land, and water. To develop the inventory, a flow model of the technical system is constructed using data on inputs and outputs. The flow model is typically illustrated with a flow chart that includes the activities that are going to be assessed in the relevant supply chain and gives a clear picture of the technical system boundaries. The input and output data needed for the construction of the model are collected for all activities within the system boundary, including from the supply chain (referred to as inputs from the technosphere).

The data must be related to the functional unit defined in the goal and scope definition. Data can be presented in tables and some interpretations can be made already at this stage. The results of the inventory is an LCI which provides information about all inputs and outputs in the form of elementary flow to and from the environment from all the unit processes involved in the study.

Inventory flows can number in the hundreds depending on the system boundary. For product LCAs at either the generic (i.e., representative industry averages) or brand-specific level, that data is typically collected through survey questionnaires. At an industry level, care has to be taken to ensure that questionnaires are completed by a representative sample of producers, leaning toward neither the best nor the worst, and fully representing any regional differences due to energy use, material sourcing or other factors. The questionnaires cover the full range of inputs and outputs, typically aiming to account for 99% of the mass of a product, 99% of the energy used in its production and any environmentally sensitive flows, even if they fall within the 1% level of inputs.

One area where data access is likely to be difficult is flows from the technosphere. The technosphere is more simply defined as the human-made world. Considered by geologists as secondary resources, these resources are in theory 100% recyclable; however, in a practical sense, the primary goal is salvage. For an LCI, these technosphere products (supply chain products) are those that have been produced by human and unfortunately those completing a questionnaire about a process which uses a human-made product as a means to an end will be unable to specify how much of a given input they use. Typically, they will not have access to data concerning inputs and outputs for previous production processes of the product. The entity undertaking the LCA must then turn to secondary sources if it does not already have that data from its own previous studies. National databases or data sets that come with LCA-practitioner tools, or that can be readily accessed, are the usual sources for that information. Care must then be taken to ensure that the secondary data source properly reflects regional or national conditions.

LCI Methods

- Process LCA.

- Economic Input Output LCA.
- Hybrid Approach.

Life Cycle Impact Assessment

Inventory analysis is followed by impact assessment. This phase of LCA is aimed at evaluating the significance of potential environmental impacts based on the LCI flow results. Classical life cycle impact assessment (LCIA) consists of the following mandatory elements:

- Selection of impact categories, category indicators, and characterization models;
- The classification stage, where the inventory parameters are sorted and assigned to specific impact categories; and
- Impact measurement, where the categorized LCI flows are characterized, using one of many possible LCIA methodologies, into common equivalence units that are then summed to provide an overall impact category total.

In many LCAs, characterization concludes the LCIA analysis; this is also the last compulsory stage according to ISO 14044:2006. However, in addition to the above mandatory LCIA steps, other optional LCIA elements – normalization, grouping, and weighting – may be conducted depending on the goal and scope of the LCA study. In normalization, the results of the impact categories from the study are usually compared with the total impacts in the region of interest, the U. S. for example. Grouping consists of sorting and possibly ranking the impact categories. During weighting, the different environmental impacts are weighted relative to each other so that they can then be summed to get a single number for the total environmental impact. ISO 14044:2006 generally advises against weighting, stating that "weighting, shall not be used in LCA studies intended to be used in comparative assertions intended to be disclosed to the public". This advice is often ignored, resulting in comparisons that can reflect a high degree of subjectivity as a result of weighting.

Life cycle impacts can also be categorized under the several phases of the development, production, use, and disposal of a product. Broadly speaking, these impacts can be divided into "First Impacts," use impacts, and end of life impacts. "First Impacts" include extraction of raw materials, manufacturing (conversion of raw materials into a product), transportation of the product to a market or site, construction/installation, and the beginning of the use or occupancy. Use impacts include physical impacts of operating the product or facility (such as energy, water, etc.), maintenance, renovation and repairs (required to continue to use the product or facility). End of life impacts include demolition and processing of waste or recyclable materials.

Interpretation

Life Cycle Interpretation is a systematic technique to identify, quantify, check, and evaluate information from the results of the life cycle inventory and the life cycle impact assessment. The results from the inventory analysis and impact assessment are summarized during the interpretation phase. The outcome of the interpretation phase is a set of conclusions and recommendations for the study. According to ISO 14040:2006, the interpretation should include:

- Identification of significant issues based on the results of the LCI and LCIA phases of an LCA;
- Evaluation of the study considering completeness, sensitivity and consistency checks; and
- Conclusions, limitations and recommendations.

A key purpose of performing life cycle interpretation is to determine the level of confidence in the final results and communicate them in a fair, complete, and accurate manner. Interpreting the results of an LCA is not as simple as "3 is better than 2, therefore Alternative A is the best choice"! Interpreting the results of an LCA starts with understanding the accuracy of the results, and ensuring they meet the goal of the study. This is accomplished by identifying the data elements that contribute significantly to each impact category, evaluating the sensitivity of these significant data elements, assessing the completeness and consistency of the study, and drawing conclusions and recommendations based on a clear understanding of how the LCA was conducted and the results were developed.

More specifically, the best alternative is the one that the LCA shows to have the least cradle-to-grave environmental negative impact on land, sea, and air resources.

LCA Uses

Based on a survey of LCA practitioners carried out in 2006 LCA is mostly used to support business strategy (18%) and R&D (18%), as input to product or process design (15%), in education (13%) and for labeling or product declarations (11%). LCA will be continuously integrated into the built environment as tools such as the European ENSLIC Building project guidelines for buildings or developed and implemented, which provide practitioners guidance on methods to implement LCI data into the planning and design process.

Major corporations all over the world are either undertaking LCA in house or commissioning studies, while governments support the development of national databases to support LCA. Of particular note is the growing use of LCA for ISO Type III labels called Environmental Product Declarations, defined as "quantified environmental data for a product with pre-set categories of parameters based on the ISO 14040 series of standards, but not excluding additional environmental information". These third-party certified LCA-based labels provide an increasingly important basis for assessing the relative environmental merits of competing products. Third-party certification plays a major role in today's industry. Independent certification can show a company's dedication to safer and environmental friendlier products to customers and NGOs.

LCA also has major roles in environmental impact assessment, integrated waste management and pollution studies. A recent study evaluated the LCA of a laboratory scale plant for oxygen enriched air production coupled with its economic evaluation in an holistic eco-design standpoint. LCA has also been used to assess the environmental impacts of pavement maintenance, repair, and rehabilitation activities.

Data Analysis

A life cycle analysis is only as valid as its data; therefore, it is crucial that data used for the completion of a life cycle analysis are accurate and current. When comparing different life cycle analyses with one another, it is crucial that equivalent data are available for both products or processes in question. If one product has a much higher availability of data, it cannot be justly compared to another product which has less detailed data.

There are two basic types of LCA data – unit process data and environmental input-output data (EIO), where the latter is based on national economic input-output data. Unit process data are derived from direct surveys of companies or plants producing the product of interest, carried out at a unit process level defined by the system boundaries for the study.

Data validity is an ongoing concern for life cycle analyses. Due to globalization and the rapid pace of research and development, new materials and manufacturing methods are continually being introduced to the market. This makes it both very important and very difficult to use up-to-date information when performing an LCA. If an LCA's conclusions are to be valid, the data must be recent; however, the data-gathering process takes time. If a product and its related processes have not undergone significant revisions since the last LCA data was collected, data validity is not a problem. However, consumer electronics such as cell phones can be redesigned as often as every 9 to 12 months, creating a need for ongoing data collection.

The life cycle considered usually consists of a number of stages including: materials extraction, processing and manufacturing, product use, and product disposal. If the most environmentally harmful of these stages can be determined, then impact on the environment can be efficiently reduced by focusing on making changes for that particular phase. For example, the most energy-intensive life phase of an airplane or car is during use due to fuel consumption. One of the most effective ways to increase fuel efficiency is to decrease vehicle weight, and thus, car and airplane manufacturers can decrease environmental impact in a significant way by replacing heavier materials with lighter ones such as aluminium or carbon fiber-reinforced elements. The reduction during the use phase should be more than enough to balance additional raw material or manufacturing cost.

Data sources are typically large databases. It is not appropriate to compare two options if different data sources have been used to source the data. Common data sources include:

- Soca,
- EuGeos' 15804-IA,
- NEEDS,
- Ecoinvent,
- PSILCA,
- ESU World Food,
- GaBi,
- ELCD,
- LC-Inventories.ch,
- Social Hotspots,
- ProBas,
- bioenergiedat,
- Agribalyse,
- USDA,
- Ökobaudat,
- Agri-footprint,
- Comprehensive Environmental Data Archive (CEDA).

Calculations for impact can then be done by hand, but it is more usual to streamline the process by using software. This can range from a simple spreadsheet, where the user enters the data manually to a fully automated program, where the user is not aware of the source data.

Variants

Cradle-to-grave

Cradle-to-grave is the full Life Cycle Assessment from resource extraction ('cradle') to use phase and disposal phase ('grave'). For example, trees produce paper, which can be recycled into low-energy production cellulose (fiberised paper) insulation, then used as an energy-saving device in the ceiling of a home for 40 years, saving 2,000 times the fossil-fuel energy used in its production. After 40 years the cellulose fibers are replaced and the old fibers are disposed of, possibly incinerated. All inputs and outputs are considered for all the phases of the life cycle.

Cradle-to-gate

Cradle-to-gate is an assessment of a *partial* product life cycle from resource extraction (*cradle*) to the factory gate (i.e., before it is transported to the consumer). The use phase and disposal phase of the product are omitted in this case. Cradle-to-gate assessments are sometimes the basis for environmental product declarations (EPD) termed business-to-business EPDs. One of the significant uses of the cradle-to-gate approach compiles the life cycle inventory (LCI) using cradle-to-gate. This allows the LCA to collect all of the impacts leading up to resources being purchased by the facility. They can then add the steps involved in their transport to plant and manufacture process to more easily produce their own cradle-to-gate values for their products.

Cradle-to-cradle or Closed Loop Production

Cradle-to-cradle is a specific kind of cradle-to-grave assessment, where the end-of-life disposal step for the product is a recycling process. It is a method used to minimize the environmental impact of products by employing sustainable production, operation, and disposal practices and aims to incorporate social responsibility into product development. From the recycling process originate new, identical products (e.g., asphalt pavement from discarded asphalt pavement, glass bottles from collected glass bottles), or different products (e.g., glass wool insulation from collected glass bottles).

Allocation of burden for products in open loop production systems presents considerable challenges for LCA. Various methods, such as the avoided burden approach have been proposed to deal with the issues involved.

Gate-to-gate

Gate-to-gate is a partial LCA looking at only one value-added process in the entire production chain. Gate-to-gate modules may also later be linked in their appropriate production chain to form a complete cradle-to-gate evaluation.

Well-to-wheel

Well-to-wheel is the specific LCA used for transport fuels and vehicles. The analysis is often broken down into stages entitled "well-to-station", or "well-to-tank", and "station-to-wheel" or "tank-to-wheel", or "plug-to-wheel". The first stage, which incorporates the feedstock or fuel production and processing and fuel delivery or energy transmission, and is called the "upstream" stage, while the stage that deals with vehicle operation itself is sometimes called the "downstream" stage. The well-to-wheel analysis is commonly used to assess total energy consumption, or the energy conversion efficiency and emissions impact of marine vessels, aircraft and motor vehicles, including their carbon footprint, and the fuels used in each of these transport modes. WtW analysis is useful for reflecting the different efficiencies and emissions of energy technologies and fuels at both the upstream and downstream stages, giving a more complete picture of real emissions.

The well-to-wheel variant has a significant input on a model developed by the Argonne National Laboratory. The Greenhouse gases, Regulated Emissions, and Energy use in Transportation (GREET) model was developed to evaluate the impacts of new fuels and vehicle technologies. The model evaluates the impacts of fuel use using a well-to-wheel evaluation while a traditional cradle-to-grave approach is used to determine the impacts from the vehicle itself. The model reports energy use, greenhouse gas emissions, and six additional pollutants: volatile organic compounds (VOCs), carbon monoxide (CO), nitrogen oxide (NO_x), particulate matter with size smaller than 10 micrometre (PM_{10}), particulate matter with size smaller than 2.5 micrometre ($PM_{2.5}$), and sulfur oxides (SO_x).

Quantitative values of greenhouse gas emissions calculated with the WTW or with the LCA method can differ, since the LCA is considering more emission sources. In example, while assessing the GHG emissions of a Battery Electric Vehicle in comparison with a conventional internal combustion engine vehicle, the WTW (accounting only the GHG for manufacturing the fuels) finds out that an electric vehicle can save the 50-60% of GHG, while an hybrid LCA-WTW method, considering also the GHG due to the manufacturing and the end of life of the battery gives GHG emission savings 10-13% lower, compared to the WTW.

Economic Input–output Life Cycle Assessment

Economic input–output LCA (EIOLCA) involves use of aggregate sector-level data on how much environmental impact can be attributed to each sector of the economy and how much each sector purchases from other sectors. Such analysis can account for long chains (for example, building an automobile requires energy, but producing energy requires vehicles, and building those vehicles requires energy, etc.), which somewhat alleviates the scoping problem of process LCA; however, EIOLCA relies on sector-level averages that may or may not be representative of the specific subset of the sector relevant to a particular product and therefore is not suitable for evaluating the environmental impacts of products. Additionally the translation of economic quantities into environmental impacts is not validated.

Ecologically based LCA

While a conventional LCA uses many of the same approaches and strategies as an Eco-LCA, the latter considers a much broader range of ecological impacts. It was designed to provide a guide to wise management of human activities by understanding the direct and indirect impacts on ecological resources and surrounding ecosystems. Developed by Ohio State University Center for resilience, Eco-LCA is a methodology that quantitatively takes into account regulating and supporting services during the life cycle of economic goods and products. In this approach services are categorized in four main groups: supporting, regulating, provisioning and cultural services.

Exergy based LCA

Exergy of a system is the maximum useful work possible during a process that brings the system into equilibrium with a heat reservoir. Wall clearly states the relation between exergy analysis and resource accounting. This intuition confirmed by DeWulf and Sciubba lead to Exergo-economic accounting and to methods specifically dedicated to LCA such as Exergetic material input per unit of service (EMIPS). The concept of material input per unit of service (MIPS) is quantified in terms of the second law of thermodynamics, allowing the calculation of both resource input and service output in exergy terms. This exergetic material input per unit of service (EMIPS) has been elaborated for transport technology. The service not only takes into account the total mass to be transported and the total distance, but also the mass per single transport and the delivery time.

Life Cycle Energy Analysis

Life cycle energy analysis (LCEA) is an approach in which all energy inputs to a product are accounted for, not only direct energy inputs during manufacture, but also all energy inputs needed to produce components, materials and services needed for the manufacturing process. An earlier term for the approach was energy analysis. With LCEA, the total life cycle energy input is established.

Energy Production

It is recognized that much energy is lost in the production of energy commodities themselves, such as nuclear energy, photovoltaic electricity or high-quality petroleum products. *Net energy content* is the energy content of the product minus energy input used during extraction and conversion, directly or indirectly. A controversial early result of LCEA claimed that manufacturing solar cells requires more energy than can be recovered in using the solar cell. The result was refuted. Another new concept that flows from life cycle assessments is energy cannibalism. Energy cannibalism refers to an effect where rapid growth of an entire energy-intensive industry creates a need for energy that uses (or cannibalizes) the energy of existing power plants. Thus during rapid growth the industry as a whole produces no energy because new energy is used to fuel the embodied energy of future power plants. Work has been undertaken in the UK to determine the life cycle energy (alongside full LCA) impacts of a number of renewable technologies.

Energy Recovery

If materials are incinerated during the disposal process, the energy released during burning can be harnessed and used for electricity production. This provides a low-impact energy source, especially when compared with coal and natural gas While incineration produces more greenhouse gas emissions than landfills, the waste plants are well-fitted with filters to minimize this negative impact. A recent study comparing energy consumption and greenhouse gas emissions from landfills (without energy recovery) against incineration (with energy recovery) found incineration to be superior in all cases except for when landfill gas is recovered for electricity production.

Criticism

It has also been argued that energy efficiency is only one consideration in deciding which alternative process to employ, and that it should not be elevated to the only criterion for determining environmental acceptability. For example, simple energy analysis does not take into account the renewability of energy flows or the toxicity of waste products;. Incorporating Dynamic LCAs of renewable energy technologies (using sensitivity analyses to project future improvements in renewable systems and their share of the power grid) may help mitigate this criticism.

In recent years, the literature on life cycle assessment of energy technology has begun to reflect the interactions between the current electrical grid and future energy technology. Some papers have focused on energy life cycle, while others have focused on carbon dioxide (CO_2) and other greenhouse gases. The essential critique given by these sources is that when considering energy technology, the growing nature of the power grid must be taken into consideration. If this is not done, a given class of energy technology may emit more CO_2 over its lifetime than it initially thought it would mitigate, with this most well documented in wind energy's case.

A problem the energy analysis method cannot resolve is that different energy forms (heat, electricity, chemical energy etc.) have different quality and value even in natural sciences, as a consequence of the two main laws of thermodynamics. A thermodynamic measure of the quality of energy is exergy. According to the first law of thermodynamics, all energy inputs should be accounted with equal weight, whereas by the second law diverse energy forms should be accounted by different values.

The conflict is resolved in one of these ways:

- Value difference between energy inputs is ignored,
- A value ratio is arbitrarily assigned (e.g., a joule of electricity is 2.6 times more valuable than a joule of heat or fuel input),
- The analysis is supplemented by economic (monetary) cost analysis,
- Exergy instead of energy can be the metric used for the life cycle analysis.

Critiques

Life cycle assessment is a powerful tool for analyzing commensurable aspects of quantifiable systems. Not every factor, however, can be reduced to a number and inserted into a model. Rigid system boundaries make accounting for changes in the system difficult. This is sometimes referred to as the boundary critique to systems thinking. The accuracy and availability of data can also contribute to inaccuracy. For instance, data from generic processes may be based on averages, unrepresentative sampling, or outdated results. Additionally, social implications of products are generally lacking in LCAs. Comparative life-cycle analysis is often used to determine a better process or product to use. However, because of aspects like differing system boundaries, different statistical information, different product uses, etc., these studies can easily be swayed in favor of one product or process over another in one study and the opposite in another study based on varying parameters and different available data. There are guidelines to help reduce such conflicts in results but the method still provides a lot of room for the researcher to decide what is important, how the product is typically manufactured, and how it is typically used.

An in-depth review of 13 LCA studies of wood and paper products found a lack of consistency in the methods and assumptions used to track carbon during the product lifecycle. A wide variety of methods and assumptions were used, leading to different and potentially contrary conclusions – particularly with regard to carbon sequestration and methane generation in landfills and with carbon accounting during forest growth and product use.

Waste Minimisation

Waste minimisation is a set of processes and practices intended to reduce the amount of waste produced. By reducing or eliminating the generation of harmful and persistent wastes, waste minimisation supports efforts to promote a more sustainable society. Waste minimisation involves redesigning products and processes and/or changing societal patterns of consumption and production.

The most environmentally resourceful, economically efficient, and cost effective way to manage waste often is to not have to address the problem in the first place. Managers see waste minimisation as a primary focus for most waste management strategies. Proper waste treatment and disposal can require a significant amount of time and resources; therefore, the benefits of waste minimisation can be considerable if carried out in an effective, safe and sustainable manner.

Traditional waste management focuses on processing waste after it is created, concentrating on re-use, recycling, and waste-to-energy conversion. Waste minimisation involves efforts to avoid creating the waste during manufacturing. To effectively implement waste minimisation the manager requires knowledge of the production process, cradle-to-grave analysis (the tracking of materials from their extraction to their return to earth) and details of the composition of the waste.

The main sources of waste vary from country to country. In the UK, most waste comes from the construction and demolition of buildings, followed by mining and quarrying, industry and commerce. Household waste constitutes a relatively small proportion of all waste. Industrial waste is often tied to requirements in the supply chain. For example, a company handling a product may insist that it should be shipped using particular packing because it fits downstream needs.

Benefits

Waste minimisation can protect the environment and often turns out to have positive economic benefits. Waste minimisation can improve:

- Efficient production practices: Waste minimisation can achieve more output of product per unit of input of raw materials.
- Economic returns: More efficient use of products means reduced costs of purchasing new materials improving the financial performance of a company.
- Public image: The environmental profile of a company is an important part of its overall reputation and waste minimisation reflects a proactive movement towards environmental protection.
- Quality of products produced: New innovation and technological practices can reduce waste generation and improve the quality of the inputs in the production phase.
- Environmental responsibility: Minimising or eliminating waste generation makes it easier to meet targets of environmental regulations, policies, and standards. The environmental impact of waste will be reduced.

Industries

In industry, using more efficient manufacturing processes and better materials generally reduces the production of waste. The application of waste minimisation techniques has led to the development of innovative and commercially successful replacement products.

Waste minimisation efforts often require investment, which is usually compensated by the savings. However, waste reduction in one part of the production process may create waste production in another part.

Processes

- Reuse of scrap material.

  Scraps can be immediately re-incorporated at the beginning of the manufacturing line so that they do not become a waste product. Many industries routinely do this; for example, paper mills return any damaged rolls to the beginning of the production line, and in the manufacture of plastic items, off-cuts and scrap are re-incorporated into new products.

- Improved quality control and process monitoring.

  Steps can be taken to ensure that the number of reject batches is kept to a minimum. This is achieved by increasing the frequency of inspection and the number of points of inspection. For example, installing automated continuous monitoring equipment can help to identify production problems at an early stage.

- Waste exchanges.

  This is where the waste product of one process becomes the raw material for a second process. Waste exchanges represent another way of reducing waste disposal volumes for waste that cannot be eliminated.

- Ship to point of use.

  This involves making deliveries of incoming raw materials or components direct to the point where they are assembled or used in the manufacturing process to minimise handling and the use of protective wrappings or enclosures.

- Zero waste.

  This is a whole systems approach that aims to eliminate waste at the source and at all points down the supply chain, with the intention of producing no waste. It is a design philosophy which emphasizes waste prevention as opposed to end of pipe waste management. Since, globally speaking, waste as such, however minimal, can never be prevented (there will always be an end-of-life even for recycled products and materials), a related goal is pollution prevention.

Product Design

Waste minimisation and resource maximisation for manufactured products can most easily be done at the design stage. Reducing the number of components used in a product or making the product easier to take apart can make it easier to be repaired or recycled at the end of its useful life.

In some cases, it may be best not to minimise the volume of raw materials used to make a product, but instead reduce the volume or toxicity of the waste created at the end of a product's life, or the environmental impact of the product's use.

Fitting the Intended Use

In this strategy, products and packages are optimally designed to meet their intended use. This applies especially to packaging materials, which should only be as durable as necessary to serve their intended purpose. On the other hand, it could be more wasteful if food, which has consumed resources and energy in its production, is damaged and spoiled because of extreme measures to reduce the use of paper, metals, glass and plastics in its packaging.

Durability

Improving product durability, such as extending a vacuum cleaner's useful life to 15 years instead of 12, can reduce waste and usually much improves resource optimisation.

But in some cases it has a negative environmental impact. If a product is too durable, its replacement with more efficient technology is likely to be delayed. Therefore, extending an older machine's useful life may place a heavier burden on the environment than scrapping it, recycling its metal and buying a new model. Similarly, older vehicles consume more fuel and produce more emissions than their modern counterparts.

Most proponents of waste minimisation consider that the way forward may be to view any manufactured product at the end of its useful life as a resource for recycling and reuse rather than waste.

Making refillable glass bottles strong enough to withstand several journeys between the consumer and the bottling plant requires making them thicker and so heavier, which increases the resources required to transport them. Since transport has a large environmental impact, careful evaluation is required of the number of return journeys bottles make. If a refillable bottle is thrown away after being refilled only several times, the resources wasted may be greater than if the bottle had been designed for a single journey.

Many choices involve trade-offs of environmental impact, and often there is insufficient information to make informed decisions.

Retail

Various aspects of business practices affect waste, such as the use of disposable tableware in restaurants.

Reusable Shopping Bags

Reusable bags are a visible form of re-use, and some stores offer a "bag credit" for reusable shopping bags, although at least one chain reversed its policy, claiming "it was just a temporary bonus". In contrast, one study suggests that a bag tax is a more effective incentive than a similar discount. (the before/after study compared a circumstance in which some stores offered a discount vs. a circumstance in which all stores applying the tax.) While there is a minor inconvenience involved, this may remedy itself, as reusable bags are generally more convenient for carrying groceries.

Households

Appropriate amounts and sizes can be chosen when purchasing goods; buying large containers of paint for a small decorating job or buying larger amounts of food than can be consumed create unnecessary waste. Also, if a pack or can is to be thrown away, any remaining contents must be removed before the container can be recycled.

Home composting, the practice of turning kitchen and garden waste into compost can be considered waste minimisation.

The resources that households use can be reduced considerably by using electricity thoughtfully (e.g. turning off lights and equipment when it is not needed) and by reducing the number of car journeys made. Individuals can reduce the amount of waste they create by buying fewer products and by buying products which last longer. Mending broken or worn items of clothing or equipment also contributes to minimising household waste. Individuals can minimise their water usage, and walk or cycle to their destination rather than using their car to save fuel and cut down emissions.

In a domestic situation, the potential for minimisation is often dictated by lifestyle. Some people may view it as wasteful to purchase new products solely to follow fashion trends when the older products are still usable. Adults working full-time have little free time, and so may have to purchase more convenient foods that require little preparation, or prefer disposable nappies if there is a baby in the family.

The amount of waste an individual produces is a small portion of all waste produced by society, and personal waste reduction can only make a small impact on overall waste volumes. Yet, influence on policy can be exerted in other areas. Increased consumer awareness of the impact and power of certain purchasing decisions allows industry and individuals to change the total resource consumption. Consumers can influence manufacturers and distributors by avoiding buying products that do not have eco-labelling, which is currently not mandatory, or choosing products that minimise the use of packaging. In the UK, PullApart combines both environmental and consumer packaging surveys, in a curbside packaging recycling classification system to minimise waste. Where reuse schemes are available, consumers can be proactive and use them.

Health-care Facilities

Health-care establishments are massive producers of waste. The major sources of health-care waste are: hospitals, laboratories and research centres, mortuary and autopsy centres, animal research and testing laboratories, blood banks and collection services, and nursing homes for the elderly.

Waste minimisation can offer many opportunities to these establishments to use fewer resources, be less wasteful and generate less hazardous waste. Good management and control practices among health-care facilities can have a significant effect on the reduction of waste generated each day.

Practices

There are many examples of more efficient practices that can encourage waste minimization in healthcare establishments and research facilities.

Source Reduction

- Purchasing reductions which ensures the selection of supplies that are less wasteful or less hazardous.
- The use of physical rather than chemical cleaning methods such as steam disinfection instead of chemical disinfection.
- Preventing the unnecessary wastage of products in nursing and cleaning activities.

Management and Control Measures at Hospital Level

- Centralized purchasing of hazardous chemicals.
- Monitoring the flow of chemicals within the health care facility from receipt as a raw material to disposal as a hazardous waste.
- The careful separation of waste matter to help minimize the quantities of hazardous waste and disposal.

Stock Management of Chemical and Pharmaceutical Products

- Frequent ordering of relatively small quantities rather than large quantities at one time.
- Using the oldest batch of a product first to avoid expiration dates and unnecessary waste.
- Using all the contents of a container containing hazardous waste.
- Checking the expiry date of all products at the time of delivery.

Environmental Management System

An Environmental Management System (EMS) is a framework that helps an organization achieve its environmental goals through consistent review, evaluation, and improvement of its environmental performance. The assumption is that this consistent review and evaluation will identify opportunities for improving and implementing the environmental performance of the organization. The EMS itself does not dictate a level of environmental performance that must be achieved; each organization's EMS is tailored to the its own individual objectives and targets.

An EMS helps an organization address its regulatory demands in a systematic and cost-effective manner. This proactive approach can help reduce the risk of non-compliance and improve health and safety practices for employees and the public. An EMS can also help address non-regulated issues, such as energy conservation, and can promote stronger operational control and employee stewardship. Basic Elements of an EMS include the following:

- Reviewing the organization's environmental goals.
- Analyzing its environmental impacts and legal requirements.
- Setting environmental objectives and targets to reduce environmental impacts and comply with legal requirements.
- Establishing programs to meet these objectives and targets.
- Monitoring and measuring progress in achieving the objectives.
- Ensuring employees' environmental awareness and competence.
- Reviewing progress of the EMS and making improvements.

Costs and Benefits of an EMS

Internal:

- Staff/manager time (represents the bulk of EMS resources expended by most organizations).
- Other employee time.

External:

- Potential consulting assistance.
- Outside training of personnel.

Potential Benefits

- Improved environmental performance.
- Enhanced compliance.

- Pollution prevention.
- Resource conservation.
- New customers/markets.
- Increased efficiency/reduced costs.
- Enhanced employee morale.
- Enhanced image with public, regulators, lenders, investors.
- Employee awareness of environmental issues and responsibilities.

EMS under ISO 14001

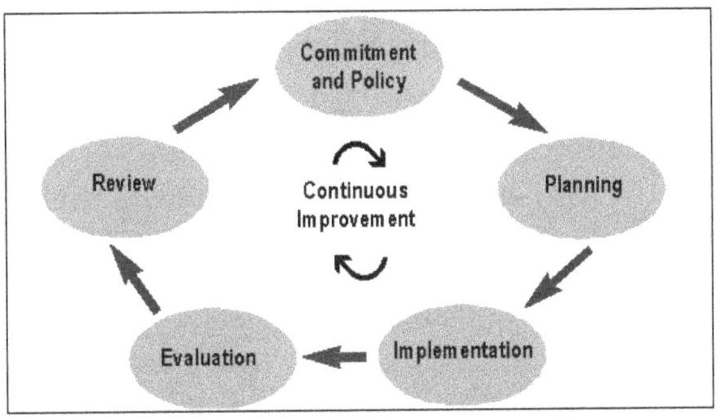

The continuous improvement cycle.

An EMS encourages an organization to continuously improve its environmental performance. The system follows a repeating cycle. The organization first commits to an environmental policy, then uses its policy as a basis for establishing a plan, which sets objectives and targets for improving environmental performance. The next step is implementation. After that, the organization evaluates its environmental performance to see whether the objectives and targets are being met. If targets are not being met, corrective action is taken. The results of this evaluation are then reviewed by top management to see if the EMS is working. Management revisits the environmental policy and sets new targets in a revised plan. The company then implements the revised plan. The cycle repeats, and continuous improvement occurs.

The most commonly used framework for an EMS is the one developed by the International Organization for Standardization (ISO) for the ISO 14001 standard EXIT. Established in 1996, this framework is the official international standard for an EMS which is based on the Plan-Do-Check-Act methodology. The five main stages of an EMS, as defined by the ISO 14001 standard EXIT, are described below:

- Commitment and Policy: Top management commits to environmental improvement and establishes the organization's environmental policy. The policy is the foundation of the EMS.

- Planning: An organization first identifies environmental aspects of its operations. Environmental aspects are those items, such as air pollutants or hazardous waste, that can have negative impacts on people and the environment. An organization then determines which aspects are significant by choosing criteria considered most important by the organization. For example, an organization may choose worker health and safety, environmental compliance, and cost as its criteria. Once significant environmental aspects are determined, an organization sets objectives and targets. An objective is an overall environmental goal (e.g., minimize use of chemical X). A target is a detailed, quantified requirement that arises from the objectives. The final part of the planning stage is devising an action plan for meeting the targets. This includes designating responsibilities, establishing a schedule, and outlining clearly defined steps to meet the targets.
- Implementation: A organization follows through with the action plan using the necessary resources (human, financial, etc.). An important component is employee training and awareness for all employees. Other steps in the implementation stage include documentation, following operating procedures, and setting up internal and external communication lines.
- Evaluation: A company monitors its operations to evaluate whether targets are being met. If not, the company takes corrective action.
- Review: Top management reviews the results of the evaluation to see if the EMS is working. Management determines whether the original environmental policy is consistent with the organization's values. The plan is then revised to optimize the effectiveness of the EMS. The review stage creates a loop of continuous improvement for a company.

Reference

- Tragedy-of-the-commons, science: britannica.com, Retrieved 3 March, 2020
- Eccleston, Charles H. (2010). Global Environmental Policy: Concepts, Principles, and Practice. Chapter 7. ISBN 978-1439847664
- OCHA. "Climate Change - Threats and Solutions". Archived from the original on 2016-04-07. Retrieved 29 June 2017
- Huesemann, Michael H., and Joyce A. Huesemann (2011). Technofix: Why Technology Won't Save Us or the Environment, Chapter 6, "Sustainability or Collapse?", New Society Publishers, ISBN 0865717044
- "ISDR : Terminology". The International Strategy for Disaster Reduction. 2004-03-31. Retrieved 2010-06-09
- Waqas Nawaz, Patrick Linke, Muammer Koç (2019). "Safety and sustainability nexus: A review and appraisal". Journal of Cleaner Production. 216: 74–87. Doi:10.1016/j.jclepro.2019.01.167

- "Curriculum - Westphal College of Media Arts & Design". Westphal College of Media Arts & Design. Retrieved 2016-11-08
- Habitat-conservation, biodiversity, marine-conservation-biology, conservation: marinebio.org, Retrieved 30 January, 2020
- Sciubba, E (2004). "From Engineering Economics to Extended Exergy Accounting: A Possible Path from Monetary to Resource-Based Costing" (PDF). Journal of Industrial Ecology. 8 (4): 19–40. Doi:10.1162/1088198043630397
- Learn-about-environmental-management-systems, ems: epa.gov, Retrieved 29 June, 2020

Natural Resource Management

3

- Environmental Resource Management
- Wildlife Management
- Sustainable Forest Management
- Planetary Management
- Natural Resource Economics
- Integrated Landscape Management
- Natural Farming
- Ecosystem-based Management
- Gis and Natural Resource Management
- Ways to Conserve Natural Resources

The management of natural resources like soil, water, land, plants and animals is referred to as natural resource management. It includes wildlife management, fisheries management, natural resource economics, integrated landscape management, etc. All these methods related to natural resource management have been thoroughly discussed in this chapter.

Natural resource management are the ways in which societies manage the supply of or access to the natural resources upon which they rely for their survival and development. Insofar as humans are fundamentally dependent on natural resources, ensuring the ongoing access to or a steady provision of natural resources has always been central to the organization of civilizations and, historically, has been organized through a range of schemes varying in degrees of formality and involvement from the central authorities.

A "natural" resource is one that is afforded by nature without human intervention; hence, the fertile lands or the minerals within them, rather than the crop that grows on them, are examples of a country's natural resources. Although what is considered a "resource" (or, for that matter, "natural") has varied over time and from one society to another, resources are, ultimately, riches provided by nature from which can be derived some form of benefit, whether material or immaterial. Under some definitions, only those natural resources that can renew themselves and whose exploitation relies on their regenerative capacities properly necessitate management. For example, oil is not usually considered a subject of natural resource management, whereas forests are. The use of nonrenewable resources is subject to regulation rather than management. The management of renewable natural resources seeks to balance the demands of exploitation with a respect for regenerative capacities.

A logging operation.

The emergence of a rational systematic management of natural resources can be traced back to the phase of accelerated industrialization of the late 19th century. In a period of unprecedented industrial growth, the pressures brought to bear on the supply of raw materials and natural resources by an unrelenting demand intensified the need to rationalize their utilization, so as to eliminate increasingly costly waste and to allocate them more efficiently. That coincided with a broader tendency toward rationalization, a general social pattern identified by the sociologist Max Weber that emerged in modern industrial societies in response to the large-scale reorganization of production and whereby goal-oriented rationality was increasingly infused into the organization of social activities. Natural resource management was born at the conjunction of rationalization and its twin process, bureaucratization, which yielded the first bureaucracies to manage nature.

Of course, there were huge variations in both the rates and degrees to which the different states became involved with questions of natural resource management. The French state, for example, took a heavy hand in forestry management as early as the 17th century, when wood became a strategic resource at a time of accelerated mercantilist (export-oriented) growth that relied primarily on maritime transportation—namely, wooden ships. Such local variations aside, overall it took a certain kind of state, the modern bureaucratic state, to steer the exploitation of natural resources toward principles of scientific management. In the United States, natural resource management was made a federal matter for the first time under the presidency of Theodore Roosevelt. At that time, principles of scientific management, which combined notions of rational management with in-depth scientific knowledge of the resource itself, were promoted by key figures such as Gifford Pinchot, who took a leading role in the U.S. government's management of forests in the 1890s and served as head of the Forest Service from its creation, in 1905, to 1910. In Europe a similar concern with rational resource exploitation transpired at about the same time. The International Council for the Exploration of the Sea, for example, provided a forum in which northern European countries could share concerns about maritime research and resources. It was effectively one of the first international conferences on a natural resource management question, and there too science was entrenched as a basis for exploitation of the seas, laying the grounds for future arrangement for the management of collective resources.

Encountering Earth's Limits

The 20th century saw natural resource management increasingly projected at a supranational level, where it was also collectivized. A first major impulse toward the internationalization of natural resource management was brought by the post-World War II context, with its pervasive spirit of cooperation on the one hand and its specific problems of food shortages on the other. Countries came together to address the issues of damaged capacities and insufficient production—in other words, insufficient use of available resources. That context yielded the Food and Agricultural Organization (FAO) in 1945, the International Whaling Commission in 1946, and, much later, the International Fund for Agricultural Development, established in 1977 to tackle problems of agricultural production in developing countries. The problem was seen to lie in the management of the resources rather than in the resources themselves. Therefore, the solution was to develop common solutions to management problems that were widely shared from one country to the next. The problem those organizations attempted to solve, in other words, was how to create international regimes that would disseminate better management solutions and thus enable each country to make better use of its resources.

In a second phase, attention shifted at a global level to Earth's resources as a whole—the seas, the air, and biodiversity—because of the realization that those essential resources, hitherto taken for granted, were in fact limited. That realization came about through two successive crises. First was the new awareness that humankind had reached a global environmental crisis, which triggered a second wave of bureaucracies to manage nature at both the national and international levels. At the international level the United Nations Environment Programme (UNEP) was established in 1972, and at the national level environmental ministries (in Europe) or agencies (in the United States) flourished in developed countries in the early 1970s. Significantly, the issue that popularized the environmental crisis was an issue of failed global resource management: the overexploitation of whales, which threatened certain species with extinction. That initial awareness was rapidly compounded by a global energy crisis that began in 1973. At that juncture natural resource management was recast as an issue of environmental governance, and it was connected to the new environmental discourses taking shape, such as sustainable development.

The issue of Earth's limits is an unsettled question, not least because of the sheer diversity of natural resources and the difficulty in assessing them scientifically. However, the remaining controversy revolves not around the idea that Earth's resources are limited but rather around its regenerative capacities, which in turn determines how those limits are to be considered. If they are relative, the question becomes one of regulating the access to a resource more stringently or of adapting the activity relying upon it so that it can use a more abundant primary resource. If the limits are absolute, then merely switching the activity from one dependency basis to another is simply insufficient, especially if such dependency continues to expand overall.

Conceptual Approaches to Natural Resource Management

Natural resource management ties in with applied concepts such as maximum sustainable yield (MSY) and optimum utilization. Every natural resource has its optimum utilization, or acceptable levels of use, which are established scientifically and according to which management authorities regulate its exploitation. Such a concept presupposes scientific knowledge as a basis for management and also a regulatory authority (whether national or international) capable of enforcing the exploitation of the resources in accord with such scientific knowledge. The MSY is a regulatory concept that translates precepts of population dynamics into a management tool. Population studies in fisheries have shown that, in a given population, when the deaths increase as a result of human harvesting (exploitation), reproduction rates usually start to rise (as if compensating for the deaths). Theoretically, that resultant surplus production can be harvested sustainably, provided that the harvest is consistently maintained under the MSY, which is specific to each population (rather than to the species as a whole). That is the peak level, beyond which the negative effect of decreasing numbers on the overall population starts to exceed the positive effect of increased reproduction rates. Subsequently, the population as a whole (and not just the harvestable surplus) begins to decline. On the other hand, maintaining exploitation levels below the MSY creates an efficient use of the resources' regenerative capacities, thus in principle enabling exploitation to continue indefinitely. The use of this tool, which was first developed in fisheries, has been extended more broadly, notably through its incorporation into the Convention on the Law of the Sea. However, it has tended to be associated with species-specific management regimes. It requires careful monitoring of population growth and overall health, and overharvesting can easily occur if the population suffers outside declines, such as disease or habitat loss, and harvesting levels are not adjusted. Additional regulations, such as limiting the harvest of females or immature individuals, can help ensure that populations are maintained at sustainable levels.

Another way to think about natural resource management is to think about the purpose of the management. The management objectives are determined by the purpose of the resource itself—as a primary resource, as a raw material or fuel, as a source of food, or as a recreational resource. Those uses fall into two broad categories, consumptive and nonconsumptive. Consumptive utilization implies a once-only form of use—that is, it refers to activities in which the resource is effectively consumed or used up, such that it cannot be utilized by another party. Hence, the possibility of future exploitation relies on the resource's ability to regenerate itself. Nonconsumptive utilization also uses the resource to generate economic value, but it does so without using up the resource itself. That category encompasses most recreational uses of natural resources. In the case of consumptive uses, management implies balancing exploitation with a respect for the resource's regenerative capacities. In nonconsumptive uses, management centres on regulating the ways in which humans interact with the resource and containing the negative effects of those interactions on the resource. In either case, management is always about resolving a tension between the potentially conflicting objectives of protection and exploitation. Sometimes the use of a resource may change over time or from one part of the globe to the next. The overexploitation of whales is a case in point: whales were initially a primary raw material and fuel in the West, until the mid-20th century, whereas today whales are considered a recreational resource in the West and a food in other areas. This coexistence of different forms of use around the same resource has generated conflicts.

Another strand in natural resource management literature focuses on the difficulties in managing collective resources—that is, resources not contained within specific territorial boundaries (such as the sea or air) or resources whose management at the local level has global repercussions, such as forests. According to one line of argument, known as the tragedy of the commons, collective resources lack the incentives inherently built into a privately owned resource to self-limit their exploitation so as to ensure that they will last. In other words, individual users tend to consume as much of the collective resource as possible before others can get to it, resulting in the overall loss of the resource for all. Also at stake in discussion of collective resources is the issue of private versus communal management, a question that goes to the heart of environmental politics.

Environmental Resource Management

Environmental resource management is the management of the interaction and impact of human societies on the environment. It is not, as the phrase might suggest, the management of the environment itself. Environmental resources management aims to ensure that ecosystem services are protected and maintained for future human generations, and also maintain ecosystem integrity through considering ethical, economic, and scientific (ecological) variables. Environmental resource management tries to identify factors affected by conflicts that rise between meeting needs and protecting resources. It is thus linked to environmental protection, sustainability and integrated landscape management.

The shrinking Aral Sea, an example of poor water resource management diverted for irrigation.

Significance

Environmental resource management is an issue of increasing concern, as reflected in its prevalence in several texts influencing global sociopolitical frameworks such as the Brundtland Commission's Our Common Future, which highlighted the integrated nature of environment and international development and the Worldwatch Institute's annual State of the World reports.

The environment determines the nature of people, animals, plants, and places around the Earth, affecting behaviour, religion, culture and economic practices.

Scope

Improved agricultural practices such as these terraces in northwest Iowa can serve to preserve soil and improve water quality.

Environmental resource management can be viewed from a variety of perspectives. It involves the management of all components of the biophysical environment, both living (biotic) and non-living (abiotic), and the relationships among all living species and their habitats. The environment also involves the relationships of the human environment, such as the social, cultural and economic environment, with the biophysical environment. The essential aspects of environmental resource management are ethical, economical, social, and technological. These underlie principles and help make decisions.

The concept of environmental determinism, probabilism and possibilism are significant in the concept of environmental resource management.

Environmental resource management covers many areas in science, including geography, biology, social sciences, political sciences, public policy, ecology, physics, chemistry, sociology, psychology, and physiology.

Aspects

Ethical

Environmental resource management strategies are intrinsically driven by conceptions of human-nature relationships. Ethical aspects involve the cultural and social issues relating to the environment, and dealing with changes to it. "All human activities take place in the context of certain types of relationships between society and the bio-physical world (the rest of nature)," and so, there is a great significance in understanding the ethical values of different groups around the world. Broadly speaking, two schools of thought exist in environmental ethics: Anthropocentrism and Ecocentrism, each influencing a broad spectrum of environmental resource management styles along a continuum. These styles perceive "different evidence, imperatives, and problems, and prescribe different solutions, strategies, technologies, roles for economic sectors, culture, governments, and ethics, etc."

Anthropocentrism

Anthropocentrism, "an inclination to evaluate reality exclusively in terms of human values," is an ethic reflected in the major interpretations of Western religions and the dominant economic paradigms of the industrialised world. Anthropocentrism looks at nature as existing solely for the benefit of humans, and as a commodity to use for the good of humanity and to improve human quality of life. Anthropocentric environmental resource management is therefore not the conservation of the environment solely for the environment's sake, but rather the conservation of the environment, and ecosystem structure, for humans' sake.

Ecocentrism

Ecocentrists believe in the intrinsic value of nature while maintaining that human beings must use and even exploit nature to survive and live. It is this fine ethical line that ecocentrists navigate between fair use and abuse. At an extreme of the ethical scale, ecocentrism includes philosophies such as ecofeminism and deep ecology, which evolved as a reaction to dominant anthropocentric paradigms. "In its current form, it is an attempt to synthesize many old and some new philosophical attitudes about the relationship between nature and human activity, with particular emphasis on ethical, social, and spiritual aspects that have been downplayed in the dominant economic worldview."

Economics

A water harvesting system collects rainwater from the Rock of Gibraltar into pipes that lead to tanks excavated inside the rock.

The economy functions within, and is dependent upon goods and services provided by natural ecosystems. The role of the environment is recognized in both classical economics and neoclassical economics theories, yet the environment was a lower priority in economic policies from 1950 to 1980 due to emphasis from policy makers on economic growth. With the prevalence of environmental problems, many economists embraced the notion that, "If environmental sustainability must coexist for economic sustainability, then the overall system must [permit] identification of an equilibrium between the environment and the economy." As such, economic policy makers began to incorporate the functions of the natural environment—or natural capital — particularly as a sink for wastes and for the provision of raw materials and amenities.

Debate continues among economists as to how to account for natural capital, specifically whether resources can be replaced through knowledge and technology, or whether the environment is a closed system that cannot be replenished and is finite. Economic models influence environmental resource management, in that management policies reflect beliefs about natural capital scarcity. For someone who believes natural capital is infinite and easily substituted, environmental management is irrelevant to the economy. For example, economic paradigms based on neoclassical models of closed economic systems are primarily concerned with resource scarcity, and thus prescribe legalizing the environment as an economic externality for an environmental resource management strategy. This approach has often been termed 'Command-and-control'. Colby has identified trends in the development of economic paradigms, among them, a shift towards more ecological economics since the 1990s.

Ecology

The juvenile fish bypass system.

Fencing separates big game from vehicles.

"The pairing of significant uncertainty about the behaviour and response of ecological systems with urgent calls for near-term action constitutes a difficult reality, and a common lament" for many environmental resource managers. Scientific analysis of the environment deals with several dimensions of ecological uncertainty. These include: structural uncertainty resulting from the misidentification, or lack of information pertaining to the relationships between ecological variables; *parameter uncertainty* referring to "uncertainty associated with parameter values that are not known precisely but can be assessed and reported in terms of the likelihood of experiencing a defined range of outcomes"; and *stochastic uncertainty* stemming from chance or unrelated factors. Adaptive management is considered a useful framework for dealing with situations of high levels of uncertainty though it is not without its detractors.

A common scientific concept and impetus behind environmental resource management is carrying capacity. Simply put, carrying capacity refers to the maximum number of organisms a particular resource can sustain. The concept of carrying capacity, whilst understood by many cultures over history, has its roots in Malthusian theory. An example is visible in the EU Water Framework Directive. However, "it is argued that Western scientific knowledge is often insufficient to deal with the full complexity of the interplay of variables in environmental resource management. These concerns have been recently addressed by a shift in environmental resource management approaches to incorporate different knowledge systems including traditional knowledge, reflected in approaches such as adaptive co-management community-based natural resource management and transitions management. among others.

Sustainability

Sustainability in environmental resource management involves managing economic, social, and ecological systems both within and outside an organizational entity so it can sustain itself and the system it exists in. In context, sustainability implies that rather than competing for endless growth on a finite planet, development improves quality of life without necessarily consuming more resources. Sustainably managing environmental resources requires organizational change that instills sustainability values that portrays these values outwardly from all levels and reinforces them to surrounding stakeholders. The end result should be a symbiotic relationship between the sustaining organization, community, and environment.

Many drivers compel environmental resource management to take sustainability issues into account. Today's economic paradigms do not protect the natural environment, yet they deepen human dependency on biodiversity and ecosystem services. Ecologically, massive environmental degradation and climate change threaten the stability of ecological systems that humanity depends on. Socially, an increasing gap between rich and poor and the global North-South divide denies many access to basic human needs, rights, and education, leading to further environmental destruction. The planet's unstable condition is caused by many anthropogenic sources. As an exceptionally powerful contributing factor to social and environmental change, the modern organisation has the potential to apply environmental resource management with sustainability principals to achieve highly effective outcomes. To achieve sustainable development with environmental resource management an organisation should work within sustainability principles, including social and environmental accountability, long-term planning; a strong, shared vision; a holistic focus; devolved and consensus decision making; broad stakeholder engagement and justice; transparency measures; trust; and flexibility.

Current Paradigm Shifts

To adjust to today's environment of quick social and ecological changes, some organizations have begun to experiment with new tools and concepts. Those that are more traditional and stick to hierarchical decision making have difficulty dealing with the demand for lateral decision making that supports effective participation. Whether it be a matter of ethics or just strategic advantage organizations are internalizing sustainability principles. Some of the world's largest and most profitable corporations are shifting to sustainable environmental resource management: Ford, Toyota, BMW, Honda, Shell, Du Port, Sta toil, Swiss Re, Hewlett-Packard, and Unilever, among others. An extensive study by the Boston Consulting Group reaching 1,560 business leaders from diverse regions, job positions, expertise in sustainability, industries, and sizes of organizations, revealed the many benefits of sustainable practice as well as its viability.

It is important to note that though sustainability of environmental resource management has improved, corporate sustainability, for one, has yet to reach the majority of global companies operating in the markets. The three major barriers to preventing organizations to shift towards sustainable practice with environmental resource management are not understanding what sustainability is; having difficulty modeling an economically viable case for the switch; and having a flawed execution plan, or a lack thereof. Therefore, the most important part of shifting an organization to adopt sustainability in environmental resource management would be to create a shared vision and understanding of what sustainability is for that particular organization, and to clarify the business case.

Tools

As with all management functions, effective management tools, standards and systems are required. An environmental management standard or system or protocol attempts to reduce environmental impact as measured by some objective criteria. The ISO 14001 standard is the most widely used standard for environmental risk management and is closely aligned to the European Eco-Management and Audit Scheme (EMAS). As a common auditing standard, the ISO 19011 standard explains how to combine this with quality management.

Other environmental management systems (EMS) tend to be based on the ISO 14001 standard and many extend it in various ways:

- The Green Dragon Environmental Management Standard is a five-level EMS designed for smaller organisations for whom ISO 14001 may be too onerous and for larger organisations who wish to implement ISO 14001 in a more manageable step-by-step approach.
- BS 8555 is a phased standard that can help smaller companies move to ISO 14001 in six manageable steps.
- The Natural Step focuses on basic sustainability criteria and helps focus engineering on reducing use of materials or energy use that is unsustainable in the long term.
- Natural Capitalism advises using accounting reform and a general biomimicry and industrial ecology approach to do the same thing.
- US Environmental Protection Agency has many further terms and standards that it defines as appropriate to large-scale EMS.
- The UN and World Bank has encouraged adopting a "natural capital" measurement and management framework.
- The European Union Eco-Management and Audit Scheme (EMAS).

Other strategies exist that rely on making simple distinctions rather than building top-down management "systems" using performance audits and full cost accounting. For instance, Ecological Intelligent Design divides products into consumables, service products or durables and unsaleables – toxic products that no one should buy, or in many cases, do not realize they are buying. By eliminating the unsaleables from the comprehensive outcome of any purchase, better environmental resource management is achieved without *systems*.

Recent successful cases have put forward the notion of *integrated management*. It shares a wider approach and stresses out the importance of interdisciplinary assessment. It is an interesting notion that might not be adaptable to all cases.

Wildlife Management

Wildlife management attempts to balance the needs of wildlife with the needs of people using the best available science. Wildlife management can include game keeping, wildlife conservation and pest control. Wildlife management draws on disciplines such as mathematics, chemistry, biology, ecology, climatology and geography to gain the best results.

Wildlife conservation aims to halt the loss in the Earth's biodiversity by taking into consideration ecological principles such as carrying capacity, disturbance and succession and environmental conditions such as physical geography, pedology and hydrology. Most wildlife biologists are concerned with the preservation and improvement of habitats although rewilding is increasingly being used. Techniques can include reforestation, pest control, nitrification and denitrification, irrigation, coppicing and hedge laying.

Game keeping is the management or control of wildlife for the well being of game and may include killing other animals which share the same niche or predators to maintain a high population of the more profitable species, such as pheasants introduced into woodland. In his 1933 book *Game Management*, Aldo Leopold, one of the pioneers of wildlife management as a science, defined it as "the art of making land produce sustained annual crops of wild game for recreational use".

Pest control is the control of real or perceived pests and can be used for the benefit of wildlife, farmers, game keepers or safety reasons. In the United States, wildlife management practices are often implemented by a governmental agency to uphold a law, such as the Endangered Species Act.

In the United Kingdom, wildlife management undertaken by several organizations including government bodies such as the Forestry Commission, Charities such as the RSPB and The Wildlife Trusts and privately hired gamekeepers and contractors. Legislation has also been passed to protect wildlife such as the Wildlife and Countryside Act 1981. The UK government also give farmers subsidies through the Countryside Stewardship Scheme to improve the conservation value of their farms.

The history of wildlife management begins with the game laws, which regulated the right to kill certain kinds of fish and wild animal (game). In Britain game laws developed out of the forest laws, which in the time of the Norman kings were very oppressive. Under William the Conqueror, it was as great a crime to kill one of the king's deer as to kill one of his subjects. A certain rank and standing, or the possession of a certain amount of property, were for a long time qualifications indispensably necessary to confer upon any one the right of pursuing and killing game.

The Game Act of 1831 protected game birds by establishing close seasons when they could not be legally taken. The act made it lawful to take game only with the provision of a game licence and provided for the appointment of gamekeepers around the country. The purposes of the law was to balance the needs for preservation and harvest and to manage both environment and populations of fish and game.

The Game Act 1831 protects game birds in England and Wales.

Early game laws were also enacted in the US; in 1839 Rhode Island closed the hunting season for white-tailed deer from May to November. Other regulations during this time focused primarily on restricting hunting. At this time, lawmakers did not consider population sizes or the need for preservation or restoration of wildlife habitats.

Emergence of Wildlife Conservation

The late 19th century saw the passage of the first pieces of wildlife conservation legislation and the establishment of the first nature conservation societies. The Sea Birds Preservation Act of 1869 was passed in Britain as the first nature protection law in the world after extensive lobbying from the Association for the Protection of Seabirds.

The Royal Society for the Protection of Birds was founded as the Plumage League in 1889 by Emily Williamson at her house in Manchester as a protest group campaigning against the use of great crested grebe and kittiwake skins and feathers in fur clothing. The group gained popularity and eventually amalgamated with the Fur and Feather League in Croydon to form the RSPB. The Society attracted growing support from the suburban middle-classes as well as support from many other influential figures, such as the ornithologist Professor Alfred Newton.

The National Trust formed in 1895 with the manifesto to "promote the permanent preservation, for the benefit of the nation, of lands to preserve (so far practicable) their natural aspect." On 1 May 1899, the Trust purchased two acres of Wicken Fen with a donation from the amateur naturalist Charles Rothschild, establishing the first nature reserve in Britain. Rothschild was a pioneer of wildlife conservation in Britain, and went on to establish many other nature reserves, such as one at Woodwalton Fen, near Huntingdon, in 1910. During his lifetime he built and managed his estate at Ashton Wold in Northamptonshire to maximise its suitability for wildlife, especially butterflies. Concerned about the loss of wildlife habitats, in 1912 he set up the Society for the Promotion of Nature Reserves, the forerunner of The Wildlife Trusts partnership.

During the society's early years, membership tended to be made up of specialist naturalists and its growth was comparatively slow. The first independent Trust was formed in Norfolk in 1926 as the Norfolk Naturalists Trust, followed in 1938 by the Pembrokeshire Bird Protection Society which after several subsequent changes of name is now the Wildlife Trust of South and West Wales and it was not until the 1940s and 1950s that more Naturalists' Trusts were formed in Yorkshire, Lincolnshire, Leicestershire and Cambridgeshire. These early Trusts tended to focus on purchasing land to establish nature reserves in the geographical areas they served.

Wildlife Management in the US

The profession of wildlife management was established in the United States in the 1920s and 1930s by Aldo Leopold and others who sought to transcend the purely restrictive policies of the previous generation of conservationists, such as anti-hunting activist William T. Hornaday. Leopold and his close associate Herbert Stoddard, who had both been trained in scientific forestry, argued that modern science and technology could be used to restore and improve wildlife habitat and thus produce abundant "crops" of ducks, deer, and other valued wild animals.

The institutional foundations of the profession of wildlife management were established in the 1930s, when Leopold was granted the first university professorship in wildlife management, University of Wisconsin, Madison), when Leopold's textbook 'Game Management' was published, when The Wildlife Society was founded, when the Journal of Wildlife Management began publishing, and when the first Cooperative Wildlife Research Units were established. Conservationists planned many projects throughout the 1940s. Some of which included the harvesting of female mammals such as deer to decrease rising populations. Others included waterfowl and wetland research. The Fish and Wildlife Management Act was put in place to urge farmers to plant food for wildlife and to provide cover for them.

In 1937, the Federal Aid in Wildlife Restoration Act (also known as the Pittman-Robertson Act) was passed in the U.S.. This law was an important advancement in the field of wildlife management. It placed a 10% tax on sales of guns and ammunition. The funds generated were then distributed to the states for use in wildlife management activities and research. This law is still in effect today.

Wildlife management grew after World War II with the help of the GI Bill and a postwar boom in recreational hunting. An important step in wildlife management in the United States national parks occurred after several years of public controversy regarding the forced reduction of the elk population in Yellowstone National Park. In 1963, United States Secretary of the Interior Stewart Udall appointed an advisory board to collect scientific data to inform future wildlife management. In a paper known as the Leopold Report, the committee observed that culling programs at other national parks had been ineffective, and recommended active management of Yellowstone's elk population.

Elk overpopulation in Yellowstone is thought by many wildlife biologists, such as Douglas Smith, to have been primarily caused by the extirpation of wolves from the park and surrounding environs. After wolves were removed, elk herds increased in population, reaching new highs during the mid-1930s. The increased number of elk apparently resulted in overgrazing in parts of Yellowstone. Park officials decided that the elk herd should be managed. For approximately thirty years, the park elk herds were culled: Each year some were captured and shipped to other locations, a certain number were killed by park rangers, and hunters were allowed to take more elk that migrated outside the park. By the late 1960s the herd populations dropped to historic lows (less than 4,000 for the Northern Range herd). This caused outrage among both conservationists and hunters. The park service stopped culling elk in 1968. The elk population then rebounded. Twenty years later there were 19,000 elk in the Northern Range herd, a historic high.

Since the tumultuous 1970s, when animal rights activists and environmentalists began to challenge some aspects of wildlife management, the profession has been overshadowed by the rise of conservation biology. Although wildlife managers remain central to the implementation of the Endangered Species Act and other wildlife conservation policies, conservation biologists have shifted the focus of conservation away from wildlife management's concern with the protection and restoration of single species and toward the maintenance of ecosystems and biodiversity.

Types of Wildlife Management

There are two general types of wildlife management:

- Manipulative management acts on a population, either changing its numbers by direct means or influencing numbers by the indirect means of altering food supply, habitat, density of predators, or prevalence of disease. This is appropriate when a population is to be harvested, or when it slides to an unacceptably low density or increases to an unacceptably high level. Such densities are inevitably the subjective view of the land owner, and may be disputed by animal welfare interests.

- Custodial management is preventive or protective. The aim is to minimize external influences on the population and its habitat. It is appropriate in a national park where one of the stated goals is to protect ecological processes. It is also appropriate for conservation of a threatened species where the threat is of external origin rather than being intrinsic to the system. Feeding of animals by visitors is generally discouraged.

Opposition

The control of wildlife through killing and hunting has been criticized by animal rights and animal welfare activists. Critics object to the real or perceived cruelty involved in some forms of wildlife management. Environmentalists have also opposed hunting where they believe it is unnecessary or will negatively affect biodiversity. Critics of game keeping note that habitat manipulation and predator control are often used to maintain artificially inflated populations of valuable game animals (including introduced exotics) without regard to the ecological integrity of the habitat.

Game keepers in the UK claim it to be necessary for wildlife conservation as the amount of countryside they look after exceeds by a factor of nine the amount in nature reserves and national parks.

Management of Hunting Seasons

Wildlife management studies, research and lobbying by interest groups help designate times of the year when certain wildlife species can be legally hunted, allowing for surplus animals to be removed. In the United States, hunting season and bag limits are determined by guidelines set by the United States Fish and Wildlife Service for migratory game such as waterfowl and other migratory gamebirds. The hunting season and bag limits for state regulated game species such as deer are usually determined by State game Commissions, which are made up of representatives from various interest groups, wildlife biologists, and researchers.

Open and closed season on deer in the UK is legislated for in the Deer Act 1991 and the Deer Act 1996.

Open Season

Open season is when wildlife is allowed to be hunted by law and is usually not during the breeding season. Hunters may be restricted by sex, age or class of animal, for instance there may be an open season for any male deer with 4 points or better on at least one side.

Limited Entry

Where the number of animals taken is to be tightly controlled, managers may have a type of lottery system called limited. Many apply, few are chosen. These hunts may still have age, sex or class restrictions.

Closed Season

Closed season is when wildlife is protected from hunting and is usually during its breeding season. Closed season is enforced by law, any hunting during closed season is punishable by law and termed as illegal hunting or poaching.

Type of Weapon Used

In wildlife management one of the conservation principles is that the weapon used for hunting should be the one that causes the least damage to the animal and is sufficiently effective so that it hits the target. Given State and Local laws, types of weapon can also vary depending on type, size, sex of game and also the geographical layout of that specific hunting area.

Sustainable Forest Management

Sustainable forest management is the management of forests according to the principles of sustainable development. Sustainable forest management has to keep the balance between three main pillars: ecological, economic and socio-cultural. Successfully achieving sustainable forest management will provide integrated benefits to all, ranging from safeguarding local livelihoods to protecting the biodiversity and ecosystems provided by forests, reducing rural poverty and mitigating some of the effects of climate change.

The "Forest Principles" adopted at The United Nations Conference on Environment and Development (UNCED) in Rio de Janeiro in 1992 captured the general international understanding of sustainable forest management at that time. A number of sets of criteria and indicators have since been developed to evaluate the achievement of SFM at the global, regional, country and management unit level. These were all attempts to codify and provide for independent assessment of the degree to which the broader objectives of sustainable forest management are being achieved in practice. In 2007, the United Nations General Assembly adopted the Non-Legally Binding Instrument on All Types of Forests. The instrument was the first of its kind, and reflected the strong international commitment to promote implementation of sustainable forest management through a new approach that brings all stakeholders together.

A definition of SFM was developed by the Ministerial Conference on the Protection of Forests in Europe (FOREST EUROPE), and has since been adopted by the Food and Agriculture Organization (FAO). It defines sustainable forest management as:

> The stewardship and use of forests and forest lands in a way, and at a rate, that maintains their biodiversity, productivity, regeneration capacity, vitality and their potential to fulfill, now and in the future, relevant ecological, economic and social functions, at local, national, and global levels, and that does not cause damage to other ecosystems.

In simpler terms, the concept can be described as the attainment of balance – balance between society's increasing demands for forest products and benefits, and the preservation of forest health and diversity. This balance is critical to the survival of forests, and to the prosperity of forest-dependent communities.

For forest managers, sustainably managing a particular forest tract means determining, in a tangible way, how to use it today to ensure similar benefits, health and productivity in the future. Forest managers must assess and integrate a wide array of sometimes conflicting factors – commercial and non-commercial values, environmental considerations, community needs, even global impact – to produce sound forest plans. In most cases, forest managers develop their forest plans in consultation with citizens, businesses, organizations and other interested parties in and around the forest tract being managed. The tools and visualization have been recently evolving for better management practices.

The Food and Agriculture Organization of the United Nations, at the request of Member States, developed and launched the Sustainable Forest Management Toolbox in 2014, an online collection of tools, best practices and examples of their application to support countries implementing sustainable forest management.

Because forests and societies are in constant flux, the desired outcome of sustainable forest management is not a fixed one. What constitutes a sustainably managed forest will change over time as values held by the public change.

Criteria and Indicators

Deforestation of native rain forest in Rio de Janeiro City for extraction of clay for civil engineering. An example of non sustainable forest management.

Criteria and indicators are tools which can be used to conceptualise, evaluate and implement sustainable forest management. Criteria define and characterize the essential elements, as well as a set of conditions or processes, by which sustainable forest management may be assessed. Periodically measured indicators reveal the direction of change with respect to each criterion.

Criteria and indicators of sustainable forest management are widely used and many countries produce national reports that assess their progress toward sustainable forest management. There are nine international and regional criteria and indicators initiatives, which collectively involve more than 150 countries. Three of the more advanced initiatives are those of the Working Group on Criteria and Indicators for the Conservation and Sustainable Management of Temperate and Boreal Forests (also called the Montréal Process), Forest Europe, and the International Tropical Timber Organization. Countries who are members of the same initiative usually agree to produce reports at the same time and using the same indicators. Within countries, at the management unit level, efforts have also been directed at developing local level criteria and indicators of sustainable forest management. The Center for International Forestry Research, the International Model Forest Network and researchers at the University of British Columbia have developed a number of tools and techniques to help forest-dependent communities develop their own local level criteria and indicators. Criteria and Indicators also form the basis of third-party forest certification programs such as the Canadian Standards Association's Sustainable Forest Management Standards and the Sustainable Forestry Initiative Standard.

There appears to be growing international consensus on the key elements of sustainable forest management. Seven common thematic areas of sustainable forest management have emerged based on the criteria of the nine ongoing regional and international criteria and indicators initiatives. The seven thematic areas are:

- Extent of forest resources.
- Biological diversity.
- Forest health and vitality.
- Productive functions and forest resources.
- Protective functions of forest resources.
- Socio-economic functions.
- Legal, policy and institutional framework.

This consensus on common thematic areas (or criteria) effectively provides a common, implicit definition of sustainable forest management. The seven thematic areas were acknowledged by the international forest community at the fourth session of the United

Nations Forum on Forests and the 16th session of the Committee on Forestry. These thematic areas have since been enshrined in the Non-Legally Binding Instrument on All Types of Forests as a reference framework for sustainable forest management to help achieve the purpose of the instrument.

On January 5, 2012, the Montréal Process, Forest Europe, the International Tropical Timber Organization, and the Food and Agriculture Organization of the United Nations, acknowledging the seven thematic areas, endorsed a joint statement of collaboration to improve global forest related data collection and reporting and avoiding the proliferation of monitoring requirements and associated reporting burdens.

Ecosystem Approach

The Ecosystem Approach has been prominent on the agenda of the Convention on Biological Diversity (CBD) since 1995. The CBD definition of the Ecosystem Approach and a set of principles for its application were developed at an expert meeting in Malawi in 1995, known as the Malawi Principles. The definition, 12 principles and 5 points of "operational guidance" were adopted by the fifth Conference of Parties (COP5) in 2000. The CBD definition is as follows:

> The ecosystem approach is a strategy for the integrated management of land, water and living resources that promotes conservation and sustainable use in an equitable way. Application of the ecosystem approach will help to reach a balance of the three objectives of the Convention. An ecosystem approach is based on the application of appropriate scientific methodologies focused on levels of biological organization, which encompasses the essential structures, processes, functions and interactions among organisms and their environment. It recognizes that humans, with their cultural diversity, are an integral component of many ecosystems.

Sustainable forest management was recognized by parties to the Convention on Biological Diversity in 2004 (Decision VII/11 of COP7) to be a concrete means of applying the Ecosystem Approach to forest ecosystems. The two concepts, sustainable forest management and the ecosystem approach, aim at promoting conservation and management practices which are environmentally, socially and economically sustainable, and which generate and maintain benefits for both present and future generations. In Europe, the MCPFE and the Council for the Pan-European Biological and Landscape Diversity Strategy (PEBLDS) jointly recognized sustainable forest management to be consistent with the Ecosystem Approach in 2006.

Independent Certification

Growing environmental awareness and consumer demand for more socially responsible businesses helped third-party forest certification emerge in the 1990s as a credible tool for communicating the environmental and social performance of forest operations.

There are many potential users of certification, including: forest managers, scientists, policy makers, investors, environmental advocates, business consumers of wood and paper, and individuals.

With third-party forest certification, an independent organization develops standards of good forest management, and independent auditors issue certificates to forest operations that comply with those standards. Forest certification verifies that forests are well-managed – as defined by a particular standard – and chain-of-custody certification tracks wood and paper products from the certified forest through processing to the point of sale.

This rise of certification led to the emergence of several different systems throughout the world. As a result, there is no single accepted forest management standard worldwide, and each system takes a somewhat different approach in defining standards for sustainable forest management.

In its 2009–2010 Forest Products Annual Market Review United Nations Economic Commission for Europe/Food and Agriculture Organization stated: "Over the years, many of the issues that previously divided the (certification) systems have become much less distinct. The largest certification systems now generally have the same structural programmatic requirements."

Third-party forest certification is an important tool for those seeking to ensure that the paper and wood products they purchase and use come from forests that are well-managed and legally harvested. Incorporating third-party certification into forest product procurement practices can be a centerpiece for comprehensive wood and paper policies that include factors such as the protection of sensitive forest values, thoughtful material selection and efficient use of products.

There are more than fifty certification standards worldwide, addressing the diversity of forest types and tenures. Globally, the two largest umbrella certification programs are:

- Programme for the Endorsement of Forest Certification (PEFC).
- Forest Stewardship Council (FSC).

The area of forest certified worldwide is growing slowly. PEFC is the world's largest forest certification system, with more than two-thirds of the total global certified area certified to its Sustainability Benchmarks.

In North America, there are three certification standards endorsed by PEFC – the Sustainable Forestry Initiative, the Canadian Standards Association's Sustainable Forest Management Standard, and the American Tree Farm System. FSC has five standards in North America – one in the United States and four in Canada.

While certification is intended as a tool to enhance forest management practices throughout the world, to date most certified forestry operations are located in Europe and North America. A significant barrier for many forest managers in developing countries is that they lack the capacity to undergo a certification audit and maintain operations to a certification standard.

Forest Governance

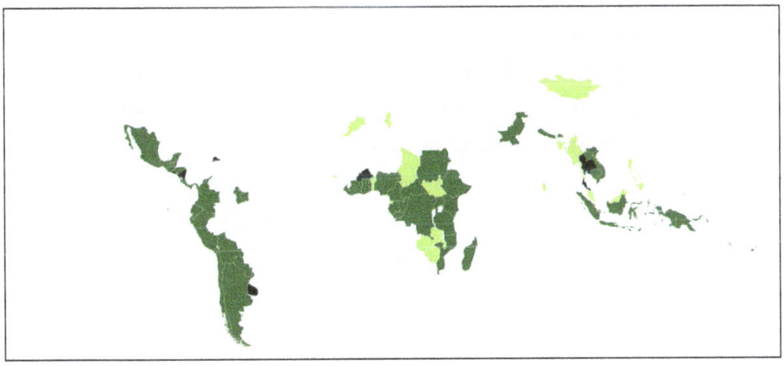

Countries participating in the UNREDD program and Forest Carbon Partnership Facility. UN-REDD participants Forest Carbon Partnership Facility participants participants in both.

Although a majority of forests continue to be owned formally by government, the effectiveness of forest governance is increasingly independent of formal ownership. Since neo-liberal ideology in the 1980s and the emanation of the climate change challenges, evidence that the state is failing to effectively manage environmental resources has emerged. Under neo-liberal regimes in the developing countries, the role of the state has diminished and the market forces have increasingly taken over the dominant socio-economic role. Though the critiques of neo-liberal policies have maintained that market forces are not only inappropriate for sustaining the environment, but are in fact a major cause of environmental destruction. Hardin's tragedy of the commons has shown that the people cannot be left to do as they wish with land or environmental resources. Thus, decentralization of management offers an alternative solution to forest governance.

The shifting of natural resource management responsibilities from central to state and local governments, where this is occurring, is usually a part of broader decentralization process. According to Rondinelli and Cheema, there are four distinct decentralization options: these are: (i) Privatization – the transfer of authority from the central government to non-governmental sectors otherwise known as market-based service provision, (ii) Delegation – centrally nominated local authority, (iii) Devolution – transfer of power to locally acceptable authority and (iv) Deconcentration – the redistribution of authority from the central government to field delegations of the central government. The major key to effective decentralization is increased broad-based participation in local-public decision making. In 2000, the World Bank report reveals that local government knows the needs and desires of their constituents better than the national government, while at the same time, it is easier to hold local leaders accountable. From the study of West African tropical forest, it is argued that the downwardly accountable and representative authorities with meaningful discretional powers are the basic institutional element of decentralization that should lead to efficiency, development and equity. This collaborates with the World Bank report in 2000 which says that decentralization should improve resource allocation, efficiency, accountability and equity "by linking the cost and benefit of local services more closely".

Many reasons point to the advocacy of decentralization of forest. (i) Integrated rural development projects often fail because they are top-down projects that did not take local people's needs and desires into account. (ii) National government sometimes have legal authority over vast forest areas that they cannot control, thus, many protected area projects result in increased biodiversity loss and greater social conflict. Within the sphere of forest management, as state earlier, the most effective option of decentralization is "devolution"-the transfer of power to locally accountable authority. However, apprehension about local governments is not unfounded. They are often short of resources, may be staffed by people with low education and are sometimes captured by local elites who promote clientelist relation rather than democratic participation. Enters and Anderson point that the result of community-based projects intended to reverse the problems of past central approaches to conservation and development have also been discouraging.

Broadly speaking, the goal of forest conservation has historically not been met when, in contrast with land use changes; driven by demand for food, fuel and profit. It is necessary to recognize and advocate for better forest governance more strongly given the importance of forest in meeting basic human needs in the future and maintaining ecosystem and biodiversity as well as addressing climate change mitigation and adaptation goal. Such advocacy must be coupled with financial incentives for government of developing countries and greater governance role for local government, civil society, private sector and NGOs on behalf of the "communities".

National Forest Funds

The development of National Forest Funds is one way to address the issue of financing sustainable forest management. National forest funds (NFFs) are dedicated financing mechanisms managed by public institutions designed to support the conservation and sustainable use of forest resources. There are 70 NFFs operating globally.

Forest Genetic Resources

Appropriate use and long-term conservation of forest genetic resources (FGR) is a part of sustainable forest management. In particular when it comes to the adaptation of forests and forest management to climate change. Genetic diversity ensures that forest trees can survive, adapt and evolve under changing environmental conditions. Genetic diversity in forests also contributes to tree vitality and to the resilience towards pests and diseases. Furthermore, FGR has a crucial role in maintaining forest biological diversity at both species and ecosystem levels.

Selecting carefully the forest reproductive material with emphasis on getting a high genetic diversity rather than aiming at producing a uniform stand of trees, is essential for sustainable use of FGR. Considering the provenance is crucial as well. For example in relation to climate change, local material may not have the genetic diversity or phenotypic plasticity to guarantee good performance under changed conditions. A different population from further away, which may have experienced selection under conditions more like those forecast for the site to be reforested, might represent a more suitable seed source.

Planetary Management

Planetary management is intentional global-scale management of Earth's biological, chemical and physical processes and cycles (water, carbon, nitrogen, sulfur, phosphorus, and others). Planetary management also includes managing humanity's influence on planetary-scale processes. Effective planetary management aims to prevent destabilisation of Earth's climate, protect biodiversity and maintain or improve human well-being. More specifically, it aims to benefit society and the global economy, and safeguard the ecosystem services upon which humanity depends – global climate, freshwater supply, food, energy, clean air, fertile soil, pollinators, and so on.

Because of the sheer complexity and enormous scope of the task, it remains to be seen whether planetary management is a feasible paradigm for maintaining global sustainability. The concept currently has defenders and critics on both sides: environmentalist David W. Orr questions whether such a task can be accomplished with human help and technology or without first examining the underlying human causes, while geographer Vaclav Smil acknowledges that "the idea of planetary management may seem preposterous to many, but at this time in history there is no rational alternative".

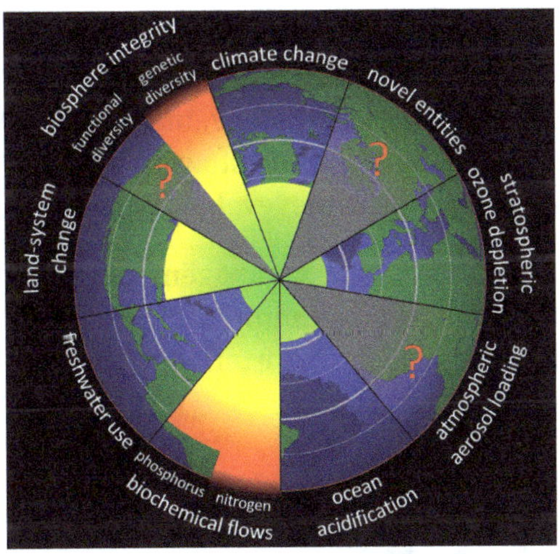

Planetary boundaries according to Rockström *et al.* 2009 and Steffen *et al.* 2015. The green areas represent human activities that are within safe margins, the yellow areas represent human activities that may or may not have exceeded safe margins, the red areas represent human activities that have exceeded safe margins, and the gray areas with red question marks represent human activities for which safe margins have not yet been determined.

The term has been around in science fiction novels since the 1970s. In 2004, the International Geosphere-Biosphere Programme published "Global Change and the Earth System, a planet under pressure." The publication's executive summary concluded: "An overall, comprehensive, internally consistent strategy for stewardship of the Earth system is required". It stated that a research goal is to define and maintain a stable equilibrium in the global environment. In 2009, the planetary boundaries concept was published in the science journal Nature. The paper identifies nine boundaries in the Earth system. Remaining within these nine boundaries, the authors suggest, may safeguard the current equilibrium.

In 2007, France called for UNEP to be replaced by a new and more powerful organization, the United Nations Environment Organization. The rationale was that UNEP's status as a programme, rather than an organization in the tradition of the World Health Organization or the World Meteorological Organization, weakened it to the extent that it was no longer fit for purpose given current knowledge of the state of Earth. The call was backed by 46 countries. Notably, the top five emitters of greenhouse gases failed to support the call.

Comparison with other Environmental Worldviews

Together with planetary management, stewardship and environmental wisdom are different ways to manage the Earth or "environmental worldviews".

In particular:

- Planetary management focus its attention on humans needs and wants, while stewardship aim at the benefit of humans, organisms and ecosystems of the Earth: in other words, planetary management considers humans as the most important species in the Earth, while stewardship and environmental wisdom consider all the species at the same level of importance.
- To accomplish its scope, planetary management relies on technology and innovation (as stewardship and spaceship-earth worldview), while environmental wisdom relies on the lesson learned from nature.

Natural Resource Economics

Natural resource economics deals with the supply, demand, and allocation of the

Earth's natural resources. One main objective of natural resource economics is to better understand the role of natural resources in the economy in order to develop more sustainable methods of managing those resources to ensure their availability to future generations. Resource economists study interactions between economic and natural systems, with the goal of developing a sustainable and efficient economy.

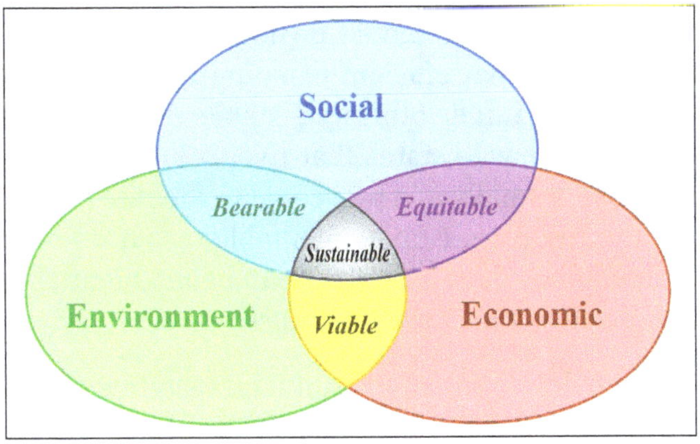

The three pillars of sustainability.

Three circles enclosed within one another showing how both economy and society are subsets of our planetary ecological system. This view is useful for correcting the misconception, sometimes drawn from the previous "three pillars" diagram, that portions of social and economic systems can exist independently from the environment.

Natural resource economics is a transdisciplinary field of academic research within economics that aims to address the connections and interdependence between human economies and natural ecosystems. Its focus is how to operate an economy within the ecological constraints of earth's natural resources. Resource economics brings together and connects different disciplines within the natural and social sciences connected to broad areas of earth science, human economics, and natural ecosystems. Economic models must be adapted to accommodate the special features of natural resource inputs. The traditional curriculum of natural resource economics emphasized fisheries models, forestry models, and minerals extraction models (i.e. fish, trees, and ore). In recent years, however, other resources, notably air, water, the global climate, and "environmental resources" in general have become increasingly important to policy-making.

Academic and policy interest has now moved beyond simply the optimal commercial exploitation of the standard trio of resources to encompass management for other objectives. For example, natural resources more broadly defined have recreational, as well as commercial values. They may also contribute to overall social welfare levels, by their mere existence.

The economics and policy area focuses on the human aspects of environmental problems. Traditional areas of environmental and natural resource economics include welfare theory, land/location use, pollution control, resource extraction, and non-market valuation, and also resource exhaustibility, sustainability, environmental management, and environmental policy. Research topics could include the environmental impacts of agriculture, transportation and urbanization, land use in poor and industrialized countries, international trade and the environment, climate change, and methodological advances in non-market valuation, to name just a few.

Hotelling's rule is a 1938 economic model of non-renewable resource management by Harold Hotelling. It shows that efficient exploitation of a nonrenewable and non-augmentable resource would, under otherwise stable economic conditions, lead to a depletion of the resource. The rule states that this would lead to a net price or "Hotelling rent" for it that rose annually at a rate equal to the rate of interest, reflecting the increasing scarcity of the resource. Nonaugmentable resources of inorganic materials (i.e. minerals) are uncommon; most resources can be augmented by recycling and by the existence and use of substitutes for the end-use products.

Vogely has stated that the development of a mineral resource occurs in five stages: (1) The current operating margin (rate of production) governed by the proportion of the reserve (resource) already depleted. (2) The intensive development margin governed by the trade-off between the rising necessary investment and quicker realization of revenue. (3) The extensive development margin in which extraction is begun of known but previously uneconomic deposits. (4) The exploration margin in which the search for new deposits (resources) is conducted and the cost per unit extracted is highly uncertain with the cost of failure having to be balanced against finding usable resources (deposits) that have marginal costs of extraction no higher than in the first three stages above. (5) The technology margin which interacts with the first four stages. The Gray-Hotelling (exhaustion) theory is a special case, since it covers only Stages 1–3 and not the far more important Stages 4 and 5.

Simon has stated that the supply of natural resources is infinite (i.e. perpetual). Furthermore, Hartwick's rule provides insight to the sustainability of welfare in an economy that uses non-renewable resources.

Perpetual Resources vs. Exhaustibility

The perpetual resource concept is a complex one because the concept of resource is complex and changes with the advent of new technology (usually more efficient recovery), new needs, and to a lesser degree with new economics (e.g. changes in prices of the material, changes in energy costs, etc.). On the one hand, a material (and its resources) can enter a time of shortage and become a strategic and critical material (an immediate exhaustibility crisis), but on the other hand a material can go out of use, its resource can proceed to being perpetual if it was not before, and then the resource can become a paleoresource when the material goes almost completely out of use (e.g. resources of arrowhead-grade flint). Some of the complexities influencing resources of a material include the extent of recyclability, the availability of suitable substitutes for the material in its end-use products, plus some other less important factors.

The Federal Government suddenly became compellingly interested in resource issues on December 7, 1941, shortly after which Japan cut the U.S. off from tin and rubber and made some other materials very difficult to obtain, such as tungsten. This was the worst case for resource availability, becoming a strategic and critical material. After the war a government stockpile of strategic and critical materials was set up, having around 100 different materials which were purchased for cash or obtained by trading off U.S. agricultural commodities for them. In the longer term, scarcity of tin later led to completely substituting aluminum foil for tin foil and polymer lined steel cans and aseptic packaging substituting for tin electroplated steel cans.

Resources change over time with technology and economics; more efficient recovery leads to a drop in the ore grade needed. The average grade of the copper ore processed has dropped from 4.0% copper in 1900 to 1.63% in 1920, 1.20% in 1940, 0.73% in 1960, 0.47% in 1980, and 0.44% in 2000.

Cobalt had been in an iffy supply status ever since the Belgian Congo (world's only significant source of cobalt) was given a hasty independence in 1960 and the cobalt-producing province seceded as Katanga, followed by several wars and insurgencies, local government removals, railroads destroyed, and nationalizations. This was topped off by an invasion of the province by Katangan rebels in 1978 that disrupted supply and transportation and caused the cobalt price to briefly triple. While the cobalt supply was disrupted and the price shot up, nickel and other substitutes were pressed into service.

Following this, the idea of a "Resource War" by the Soviets became popular. Rather than the chaos that resulted from the Zairean cobalt situation, this would be planned, a strategy designed to destroy economic activity outside the Soviet bloc by the acquisition of vital resources by noneconomic means (military?) outside the Soviet bloc (Third World?), then withholding these minerals from the West.

An important way of getting around a cobalt situation or a "Resource War" situation is to use substitutes for a material in its end-uses. Some criteria for a satisfactory substitute are (1) ready availability domestically in adequate quantities or availability from contiguous nations, or possibly from overseas allies, (2) possessing physical and chemical properties, performance, and longevity comparable to the material of first choice, (3) well-established and known behavior and properties particularly as a component in exotic alloys, and (4) an ability for processing and fabrication with minimal changes in existing technology, capital plant, and processing and fabricating facilities. Some suggested substitutions were alunite for bauxite to make alumina, molybdenum and nickel for cobalt, and aluminum alloy automobile radiators for copper alloy automobile radiators. Materials can be eliminated without material substitutes, for example by using discharges of high tension electricity to shape hard objects that were formerly shaped by mineral abrasives, giving superior performance at lower cost, or by using computers/satellites to replace copper wire (land lines).

An important way of replacing a resource is by synthesis, for example, industrial diamonds and many kinds of graphite, although a certain kind of graphite could be almost replaced by a recycled product. Most graphite is synthetic, for example, graphite electrodes, graphite fiber, graphite shapes (machined or unmachined), and graphite powder.

Another way of replacing or extending a resource is by recycling the material desired from scrap or waste. This depends on whether or not the material is dissipated or is available as a no longer usable durable product. Reclamation of the durable product depends on its resistance to chemical and physical breakdown, quantities available, price of availability, and the ease of extraction from the original product. For example, bismuth in stomach medicine is hopelessly scattered (dissipated) and therefore impossible to recover, while bismuth alloys can be easily recovered and recycled. A good example where recycling makes a big difference is the resource availability situation for graphite, where flake graphite can be recovered from a renewable resource called kish, a steelmaking waste created when carbon separates out as graphite within the kish from the molten metal along with slag. After it is cold, the kish can be processed.

Several other kinds of resources need to be introduced. If strategic and critical materials are the worst case for resources, unless mitigated by substitution and recycling, one of the best is an abundant resource. An abundant resource is one whose material has so far found little use, such as using high-aluminous clays or anorthosite to produce alumina, and magnesium before it was recovered from seawater. An abundant resource is quite similar to a perpetual resource. The reserve base is the part of an identified resource that has a reasonable potential for becoming economically available at a time beyond when currently proven technology and current economics are in operation. Identified resources are those whose location, grade, quality, and quantity are known or estimated from specific geologic evidence. Reserves are that part of the reserve base that can be economically extracted at the time of determination; reserves should not be used as a surrogate for resources because they are often distorted by taxation or the owning firm's public relations needs.

Comprehensive Natural Resource Models

Harrison Brown and associates stated that humanity will process lower and lower grade "ore". Iron will come from low-grade iron-bearing material such as raw rock from anywhere in an iron formation, not much different from the input used to make taconite pellets in North America and elsewhere today. As coking coal reserves decline, pig iron and steel production will use non-coke-using processes (i.e. electric steel). The aluminum industry could shift from using bauxite to using anorthosite and clay. Magnesium metal and magnesia consumption (i.e. in refractories), currently obtained from seawater, will increase. Sulfur will be obtained from pyrites, then gypsum or anhydrite. Metals such as copper, zinc, nickel, and lead will be obtained from manganese nodules or the Phosphoria formation (sic!). These changes could occur irregularly in different parts of the world. While Europe and North America might use anorthosite or clay as raw material for aluminum, other parts of the world might use bauxite, and while North America might use taconite, Brazil might use iron ore. New materials will appear, the result of technological advances, some acting as substitutes and some with new properties. Recycling will become more common and more efficient. Ultimately, minerals and metals will be obtained by processing "average" rock. Rock, 100 tonnes of "average" igneous rock, will yield eight tonnes of aluminum, five tonnes of iron, and 0.6 tonnes of titanium.

The USGS model based on crustal abundance data and the reserve-abundance relationship of McKelvey, is applied to several metals in the Earth's crust (worldwide) and in the U.S. crust. The potential currently recoverable (present technology, economy) resources that come closest to the McKelvey relationship are those that have been sought for the longest time, such as copper, zinc, lead, silver, gold and molybdenum. Metals that do not follow the McKelvey relationship are ones that are byproducts (of major metals) or haven't been vital to the economy until recently (titanium, aluminum to a lesser degree). Bismuth is an example of a byproduct metal that doesn't follow the relationship very well; the 3% lead reserves in the western U.S. would have only 100 ppm bismuth, clearly too low-grade for a bismuth reserve. The world recoverable resource potential is 2,120 million tonnes for copper, 2,590 million tonnes for nickel, 3,400 million tonnes for zinc, 3,519 BILLION tonnes for aluminum, and 2,035 BILLION tonnes for iron.

Diverse authors have further contributions. Some think the number of substitutes is almost infinite, particularly with the flow of new materials from the chemical industry; identical end products can be made from different materials and starting points. Plastics can be good electrical conductors. Since all materials are 100 times weaker than they theoretically should be, it ought to be possible to eliminate areas of dislocations and greatly strengthen them, enabling lesser quantities to be used. To summarize, "mining" companies will have more and more diverse products, the world economy is moving away from materials towards services, and the population seems to be levelling, all of which implies a lessening of demand growth for materials; much of the materials will be recovered from somewhat uncommon rocks, there will be much more coproducts and byproducts from a given operation, and more trade in minerals and materials.

Trend towards Perpetual Resources

As radical new technology impacts the materials and minerals world more and more powerfully, the materials used are more and more likely to have perpetual resources. There are already more and more materials that have perpetual resources and less and less materials that have nonrenewable resources or are strategic and critical materials.

Some materials that have perpetual resources such as salt,stone, magnesium, and common clay were mentioned previously. Thanks to new technology, synthetic diamonds were added to the list of perpetual resources, since they can be easily made from a lump of another form of carbon. Synthetic graphite, is made in large quantities (graphite electrodes, graphite fiber) from carbon precursors such as petroleum coke or a textile fiber. A firm named Liquidmetal Technologies, Inc. is utilizing the removal of dislocations in a material with a technique that overcomes performance limitations caused by inherent weaknesses in the crystal atomic structure. It makes amorphous metal alloys, which retain a random atomic structure when the hot metal solidifies, rather than the crystalline atomic structure (with dislocations) that normally forms when hot metal solidifies. These amorphous alloys have much better performance properties than usual; for example, their zirconium-titanium Liquidmetal alloys are 250% stronger than a standard titanium alloy. The Liquidmetal alloys can supplant many high performance alloys.

Exploration of the ocean bottom in the last fifty years revealed manganese nodules and phosphate nodules in many locations. More recently, polymetallic sulfide deposits have been discovered and polymetallic sulfide "black muds" are being presently deposited from "black smokers" The cobalt scarcity situation of 1978 has a new option now: recover it from manganese nodules. A Korean firm plans to start developing a manganese nodule recovery operation in 2010; the manganese nodules recovered would average 27% to 30% manganese, 1.25% to 1.5% nickel, 1% to 1.4% copper, and 0.2% to 0.25% cobalt (commercial grade) Nautilus Minerals Ltd. is planning to recover commercial grade material averaging 29.9% zinc, 2.3% lead, and 0.5% copper from massive ocean-bottom polymetallic sulfide deposits using an underwater vacuum cleaner-like device that combines some current technologies in a new way. Partnering with Nautilus are Tech Cominco Ltd. and Anglo-American Ltd., world-leading international firms.

There are also other robot mining techniques that could be applied under the ocean. Rio Tinto is using satellite links to allow workers 1500 kilometers away to operate drilling rigs, load cargo, dig out ore and dump it on conveyor belts, and place explosives to subsequently blast rock and earth. The firm can keep workers out of danger this way, and also use fewer workers. Such technology reduces costs and offsets declines in metal content of ore reserves. Thus a variety of minerals and metals are obtainable from unconventional sources with resources available in huge quantities.

Finally, what is a perpetual resource? The ASTM definition for a perpetual resource is "one that is virtually inexhaustible on a human time-scale". Examples given include solar energy, tidal energy, and wind energy, to which should be added salt, stone, magnesium, diamonds, and other materials mentioned above. A study on the biogeophysical aspects of sustainability came up with a rule of prudent practice that a resource stock should last 700 years to achieve sustainability or become a perpetual resource, or for a worse case, 350 years.

If a resource lasting 700 or more years is perpetual, one that lasts 350 to 700 years can be called an abundant resource, and is so defined here. How long the material can be recovered from its resource depends on human need and changes in technology from extraction through the life cycle of the product to final disposal, plus recyclability of the material and availability of satisfactory substitutes. Specifically, this shows that exhaustibility does not occur until these factors weaken and play out: the availability of substitutes, the extent of recycling and its feasibility, more efficient manufacturing of the final consumer product, more durable and longer-lasting consumer products, and even a number of other factors.

Transitioning: Perpetual Resources to Paleoresources

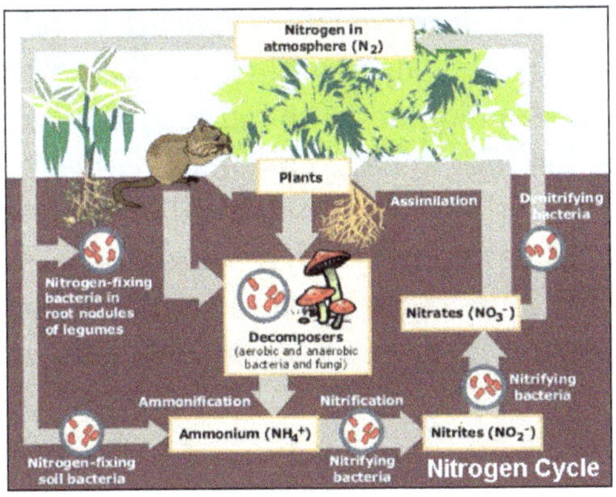

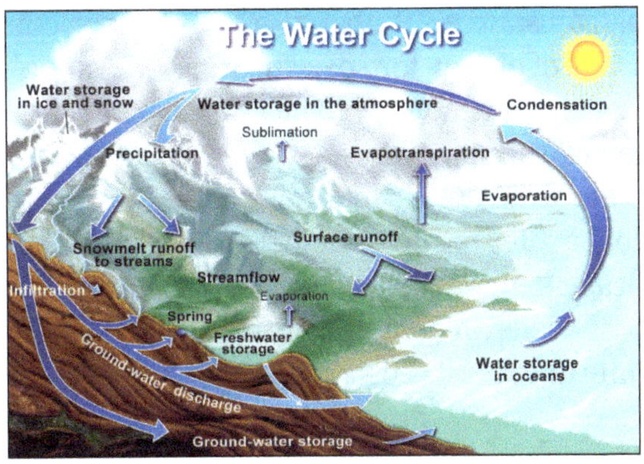

Perpetual resources can transition to being a paleoresource. A paleoresource is one that has little or no demand for the material extracted from it; an obsolescent material, humans no longer need it. The classic paleoresource is an arrowhead-grade flint resource; no one makes flint arrowheads or spearheads anymore—making a sharpened piece of scrap steel and using it is much simpler. Obsolescent products include tin cans, tin foil, the schoolhouse slate blackboard, and radium in medical technology. Radium has been replaced by much cheaper cobalt-60 and other radioisotopes in radiation treatment. Noncorroding lead as a cable covering has been replaced by plastics.

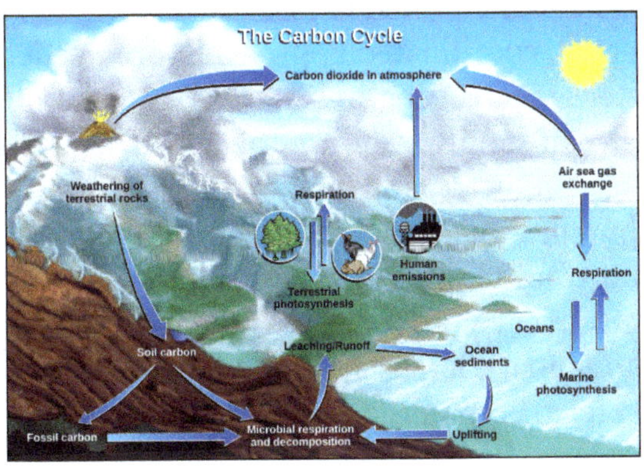

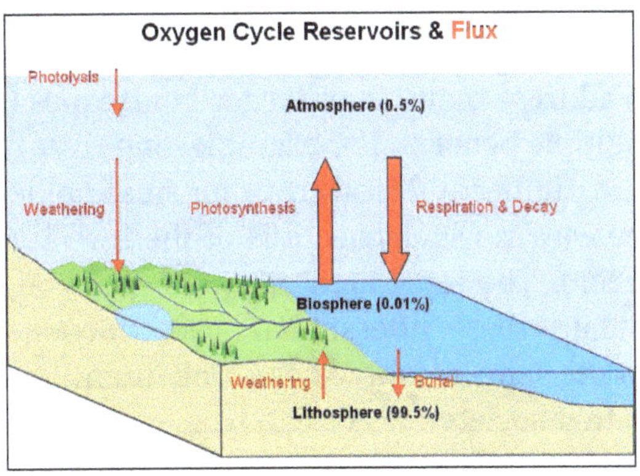

Pennsylvania anthracite is another material where the trend towards obsolescence and becoming a paleoresource can be shown statistically. Production of anthracite was 70.4 million tonnes in 1905, 49.8 million tonnes in 1945, 13.5 million tonnes in 1965, 4.3 million tonnes in 1985, and 1.5 million tonnes in 2005. The amount used per person was 84 kg per person in 1905, 7.1 kg in 1965, and 0.8 kg in 2005. Compare this to the USGS anthracite reserves of 18.6 billion tonnes and total resources of 79 billion tonnes; the anthracite demand has dropped so much that these resources are more than perpetual.

Since anthracite resources are so far into the perpetual resource range and demand for anthracite has dropped so far, is it possible to see how anthracite might become a paleoresource? Probably by customers continuing to disappear (i.e. convert to other kinds of energy for space heating), the supply network atrophy as anthracite coal dealers can't retain enough business to cover costs and close, and mines with too small a volume to cover costs also close. This is a mutually reinforcing process: customers convert to other forms of cleaner energy that produce less pollution and carbon dioxide, then the coal dealer has to close because of lack of enough sales volume to cover costs. The coal dealer's other customers are then forced to convert unless they can find another nearby coal dealer. Finally the anthracite mine closes because it doesn't have enough sales volume to cover its costs.

Integrated Landscape Management

Integrated landscape management is a way of managing a landscape that brings together multiple stakeholders, who collaborate to integrate policy and practice for their different land use objectives, with the purpose of achieving sustainable landscapes.

Integrated landscape management is one approach to addressing the major global challenges of poverty, food security, climate change, water scarcity, deforestation and loss of biodiversity at the local level. Proponents of integrated landscape management argue that as these challenges are interconnected, coordinated approaches are needed to address them, in order for landscapes (heterogenous geographic areas) to generate multiple benefits. For example, one river basin can supply water for towns and agriculture, timber and food crops for smallholders and industry, and habitat for biodiversity; the way in which each one of these sectors pursues its goals can have impacts on the others. The integrated approach goes beyond traditional sector-based practices that manage these different land uses independently of each other, even where they depend on the same resource base. The intention is to manage landscapes in a joined-up way, so that society's needs can be met in the short term, and in the long term.

Integrated landscape management is increasingly recognised and taken up by intergovernmental bodies, government initiatives, research institutes, and some of the world's largest conservation NGOs, resulting in an increase in the number of examples of the approach in practice. However, barriers to uptake include difficulties in monitoring integrated landscape management and the proliferation of definitions and terms relating to it.

A plethora of overlapping ideas and definitions have emerged, defining landscapes, landscape management, and landscape approaches from the perspectives of different fields. Despite differences in definitions, many terms can be seen as entry points for landscape approaches. One set of definitions (developed by a consortium of research groups and NGOs) that aims to create a stepwise progression through landscapes, sustainable landscapes, integrated landscape management and landscape approaches is as follows:

"A landscape is a socio-ecological system that consists of natural and human-modified ecosystems, and which is influenced by distinct ecological, historical, economic and socio-cultural processes and activities."

"A sustainable landscape helps to meet the principles of sustainable development as defined in the UN Sustainable Development Goals. These are landscapes that can meet the needs of the present without compromising the ability of future generations to meet their own needs."

"Integrated landscape management is a way of managing the landscape that involves collaboration among multiple stakeholders, with the purpose of achieving sustainable landscapes."

"The governance structure, size and scope, and number and type of stakeholders involved (private sector, civil society, government) can vary. The level of cooperation also varies, from information sharing and consultation, to more formal models with shared decision-making and joint implementation."

"A landscape approach is a conceptual framework whereby stakeholders in a landscape aim to reconcile competing social, economic and environmental objectives. It seeks to move away from the often-unsustainable sectoral approach to land management. A landscape approach aims to ensure the realisation of local level needs and action (i.e. the interests of different stakeholders within the landscape), while also considering goals and outcomes important to stakeholders outside the landscape, such as national governments or the international community."

Development and Uptake of the Approach

Efforts to develop and implement concepts that integrate social and economic development, and biodiversity conservation are not new. Integrated landscape management has evolved through multiple iterations and alongside other concepts – from watershed management to landscape ecology and from landscape-scale conservation to indigenous biocultural territorial development. It has much in common with the concept of the water-energy-food nexus, and draws on the ecosystem approach (the primary framework for action under the Convention on Biological Diversity). There are particular parallels with 'climate-smart agriculture', which also aims for landscapes to provide multiple benefits in terms of food security, rural livelihoods, and climate change adaptation and mitigation.

The interest in landscape approaches has increased in recognition of the connections between the global challenges of poverty, food security, climate change, water scarcity, deforestation and loss of biodiversity. For example, lifting people out of poverty can increase consumption and drive climate change; increasing agricultural production can exacerbate water scarcity and drive habitat loss.

Integrated landscape management has been adopted at the intergovernmental level by the UN Environment Programme, which states that "UNEP champions the landscape approach de facto as it embodies the main elements of integrated ecosystem management"; UNEP puts the approach into practice through a multinational project of Ecosystem Management of Productive Landscapes. The Convention on Biological Diversity also recognises the approach: its scientific committee has compiled information on improving sustainable use of biodiversity in a landscape perspective, and stated that "As sustainability can only be achieved at an appropriate spatial and temporal context, the landscape level is arguably the most important spatial scale to improve and assess the sustainable management of agricultural and forest ecosystems".

Landscape approaches are also being taken up by national governments (for example, to inform international cooperation for the Greater Mekong Subregion, and as a foundational principle of Indonesia's climate change commitments); by international research bodies (for example, the Center for International Forestry Research, which convenes the Global Landscapes Forum); and conservation NGOs (for example, the Worldwide Fund for Nature, Global Canopy Programme, The Nature Conservancy, IDH - The Sustainable Trade Initiative, and Eco Agriculture Partners, which collaborated on a book on integrated landscape management).

Relevance to International Commitments

Proponents of integrated landscape management argued that it is well-suited to address complex global challenges, such as those that are the focus of the Sustainable Development Goals and policies for Reducing emissions from deforestation and forest degradation (REDD+).

In terms of the Sustainable Development Goals, researchers have drawn parallels between the interdependent nature of these Goals, and the perceived potential for integrated landscape management to maximise collaboration in planning, policy development and action across multiple sectors at the landscape scale. Many of the Sustainable Development Goals have potential synergies and trade-offs (conflicts): proponents therefore argue that a sectoral approach may not be effective, whereas landscape approaches provide a potential framework to manage these interlinked challenges. To give more specific examples, increased areas of irrigated agricultural land to meet Goal 2 (ending hunger) could have adverse impacts on Goal 15 (terrestrial ecosystems) or on Goal 6 (sustainable water management). Landscape approaches are intended to systematically consider and include diverse stakeholders and sectors, and thus create a mechanism to achieve the multiple objectives of the Sustainable Development Goals – for example, working across the catchment area of a river to enhance agricultural productivity, flood defence, biodiversity and carbon storage, while also strengthened institutions for rural communities.

In terms of climate change and REDD+, unlike other sectors that emit greenhouse gases, the land use sector (including agriculture and forestry) has the potential to mitigate climate change by actively removing greenhouse gases, in addition to simply reducing emissions, especially through reforestation and landscape restoration. This sector accounts for around 24% of anthropogenic emissions - and a much greater proportion of emissions in many developing countries, where land use represents the largest potential for climate change mitigation. There are many interactions between climate change and agriculture so the case for integrated landscape management is again based on the need for landscape managers to consider multiple issues and achieve multiple objectives simultaneously (i.e. production of food and climate mitigation). Advocates of integrated landscape management argue that 'climate-smart agriculture' and implementation of the Cancun REDD+ safeguards can draw on integrated landscape management; for instance, multi-stakeholder planning, appropriate landscape governance and resource tenure, investment in the landscape, and monitoring social and climate goals at different scales.

Principles and Elements

Many ways of implementing integrated landscape management have been identified. Two long-term processes by two coalitions of researchers and NGOs developed ten principles for integrated landscape management, and five elements of integrated landscape management.

The Ten Principles of Landscape Approach include the need for continual learning and adaptive management, which incorporates participatory monitoring; the expectation that actions take place at multiple scales and that landscapes are multifunctional (e.g. supplying both goods, such as timber and food, and services, such as water and biodiversity protection); and although there will be multiple stakeholders, they nonetheless have a common concern about the landscape, negotiate change with each other in a transparent way, and their rights and responsibilities are clear or become clear through the process.

The Five Elements of Integrated Landscape Management describe the implementation cycle of a landscape approach, namely that: "interested stakeholders in the landscape come together for cooperative dialogue and action in a multi-stakeholder platform

(1); they undertake a systematic process to exchange information and discuss perspectives to achieve a shared understanding of the landscape conditions, challenges and opportunities (2); this enables collaborative planning to develop an agreed action plan (3); stakeholders then implement the plan, with attention to maintaining collaborative commitments (4); and finally, monitoring for adaptive management and accountabili-ty feeds into subsequent rounds of dialogue and the design of new collaborative action (5)". These five elements can be presented graphically in a circle, similar to the project management cycle.

Risks and Barriers

Despite many examples of implementation, and the interest of many public and private sector organisations, there are various barriers and risks around landscape approaches.

The first is terminology confusion: the variety of definitions creates the potential for not only confusion, but also resistance to engage with an approach that has emerged, often somewhat independently, from different fields. As stated by Scherr et al.: "People are talking about the same thing without realizing it. This can lead to fragmentation of knowledge, unnecessary re-invention of ideas and practices, and inability to mobilize action at scale. For policymakers, this rich diversity is often simply overwhelming: they receive confusing messages as to what an enabling policy environment to address the full set of landscape values would look like."

These risks are far from unique to landscape approaches: since the 1970s it has been recognised that the constant emergence of new terms and labels from efforts to improve and integrate policy and practice for rural areas of developing countries can be harmful if they obscure problems or promote rhetoric at the expense of action. The fact that landscapes approaches develop from, and aim to integrate, a wide variety of concepts and sectors, makes it vulnerable to overlapping definitions and parallel conceptual development. Like other approaches to conservation, it could be seen as a fad.

Further barriers to implementation have been identified. As has been the case with other integrated approaches, time lags result from the fact that substantial time and resources are invested both in developing concepts and planning projects, while resources are less adequate for implementation, and more limited still for monitoring the outcomes. Difficulties in establishing objectives, operating norms and funding that effectively bridge different sectors can lead to operating silos(where each sector pursues its goals without giving consideration to the others), whether at the level of local actors and institutions within a landscape, national ministries, or donors. Working across sectors at the landscape scale requires a range of skills, different from those traditionally used by conservation organisations. Landscape approaches rely on the engagement of multiple stakeholders, which can lead to problems of Internal/external engagement, if engagement is reduced to a box-ticking exercise or inaccessible to community participants, or powerful players such as government agencies or large companies can be deterred by a perception that there is too much participatory discussion at the expense of efficient decision-making The least developed area of landscape approaches is monitoring, to understand whether the ambitious, multiple objectives of the approach have been achieved.

Examples:

Landscape approaches are collated and analysed globally or regionally, for example from Africa and Latin America. International organisations have developed programmes that apply the approach to multiple locations worldwide (for example, the International Union for Conservation of Nature Livelihoods and Landscapes Strategies programme worked in 27 landscapes in 23 different countries). Some specific examples of implementation in individual landscapes are as follows.

An example of an indigenous producer movement that is managing a multi-functional landscape is from the Potato Park in Pisac, Peru, which is an Indigenous Biocultural Heritage Territory, where indigenous communities protect the ecological and cultural diversity of the 12000ha landscape.

An example of cooperation between very different actors is from the landscape of the Doi Mae Salong watershed in North-west Thailand, a designated Military Reserved Area under the control of the Royal Thai Armed Forces. Reforestation activities, to address degradation caused by unsustainable land uses, led to tension with local hill tribes. In response, a multi-stakeholder dialogue was used to reach agreement on land rights and use of different parts of the reserve.

Landscape near the Doi Mae Salong mountain.

The Mount Kailash region is of particular religious and ecological importance, as it is here that the Indus River, the Karnali River (a major tributary of the Ganges River), the Brahmaputra River and the Sutlej river systems originate. With assistance from the International Centre for Integrated Mountain Development (ICIMOD), the three surrounding countries (China, India and Nepal) developed an integrated management approach to the different conservation and development issues within this landscape.

In the Volta River basin, landscape modelling illustrates the cause and effect of alternative scenarios for the riparian buffer, and is used to inform land-use decisions. Six countries in West Africa are using the 'Mapping Ecosystems Services to Human well-being' (MESH) toolkit to assess strategies for conserving hydrological ecosystem services and meeting national SDG commitments.

International conservation NGOs are applying the approach using REDD+ financing, for example Conservation International is using carbon financing from Disney in the 182,000-hectare Alto Mayo Protection Forest, in the San Martín region of Peru. Smallholders sign up to REDD+ agreements, which provide incentives for them to reduce the rate of local deforestation; independent verification reported improvements in living conditions, reduced deforestation, and positive impacts on biodiversity.

Natural Farming

Masanobu Fukuoka, originator of the natural farming method.

Natural farming is an ecological farming approach established by Masanobu Fukuoka, a Japanese farmer and philosopher, introduced in his 1975 book *The One-Straw Revolution*. Fukuoka described his way of farming as (*shizen nōhō*) in Japanese. It is also referred to as "the Fukuoka Method", "the natural way of farming" or "do-nothing farming". The title refers not to lack of effort, but to the avoidance of manufactured inputs and equipment. Natural farming is related to fertility farming, organic farming, sustainable agriculture, agroecology, agroforestry, ecoagriculture and permaculture, but should be distinguished from biodynamic agriculture.

The system works along with the natural biodiversity of each farmed area, encouraging the complexity of living organisms—both plant and animal—that shape each particular ecosystem to thrive along with food plants. Fukuoka saw farming both as a means of producing food and as an aesthetic or spiritual approach to life, the ultimate goal of which was, "the cultivation and perfection of human beings". He suggested that farmers could benefit from closely observing local conditions. Natural farming is a closed system, one that demands no human-supplied inputs and mimics nature.

Fukuoka's ideas radically challenged conventions that are core to modern agro-industries; instead of promoting importation of nutrients and chemicals, he suggested an approach that takes advantage of the local environment. Although natural farming is considered a subset of organic farming, it differs greatly from conventional organic farming, which Fukuoka considered to be another modern technique that disturbs nature.

Fukuoka claimed that his approach prevents water pollution, biodiversity loss and soil erosion, while providing ample amounts of food.

In principal, practitioners of natural farming maintain that it is not a *technique* but a *view*, or a way of seeing ourselves as a part of nature, rather than separate from or above it. Accordingly, the methods themselves vary widely depending on culture and local conditions.

Rather than offering a structured method, Fukuoka distilled the natural farming mindset into five principles:

- No tillage.
- No fertilizer.
- No pesticides or herbicides.
- No weeding.
- No pruning.

A young man helps harvest rice by hand at a natural farm, in this production still from the film "Final Straw: Food, Earth, Happiness."

Though many of his plant varieties and practices relate specifically to Japan and even to local conditions in subtropical western Shikoku, his philosophy and the governing principles of his farming systems have been applied widely around the world, from Africa to the temperate northern hemisphere.

Principally, natural farming minimises human labour and adopts, as closely as practical, nature's production of foods such as rice, barley, daikon or citrus in biodiverse agricultural ecosystems. Without plowing, seeds germinate well on the surface if site conditions meet the needs of the seeds placed there. Fukuoka used the presence of spiders in his fields as a key performance indicator of sustainability.

Fukuoka specifies that the ground remain covered by weeds, white clover, alfalfa, herbaceous legumes, and sometimes deliberately sown herbaceous plants. Ground cover is present along with grain, vegetable crops and orchards. Chickens run free in orchards and ducks and carp populate rice fields.

Periodically ground layer plants including weeds may be cut and left on the surface, returning their nutrients to the soil, while suppressing weed growth. This also facilitates the sowing of seeds in the same area because the dense ground layer hides the seeds from animals such as birds.

For summer rice and winter barley grain crops, ground cover enhances nitrogen fixation. Straw from the previous crop mulches the topsoil. Each grain crop is sown before the previous one is harvested by broadcasting the seed among the standing crop. Later, this method was reduced to a single direct seeding of clover, barley and rice over the standing heads of rice. The result is a denser crop of smaller, but highly productive and stronger plants.

Fukuoka's practice and philosophy emphasised small scale operation and challenged the need for mechanised farming techniques for high productivity, efficiency and economies of scale. While his family's farm was larger than the Japanese average, he used one field of grain crops as a small-scale example of his system.

Climax Ecosystems

In ecology, climax ecosystems are mature ecosystems that have reached a high degree of stability, productivity and diversity. Natural farmers attempt to mimic those virtues, creating a comparable climax ecosystem, and employ advanced techniques such as intercropping, companion planting and integrated pest management.

No-till

Natural farming recognizes soils as a fundamental natural asset. Ancient soils possess physical and chemical attributes that render them capable of generating and supporting life abundance. It can be argued that tilling actually degrades the delicate balance of a climax soil:

- Tilling may destroy crucial physical characteristics of a soil such as *water suction*, its ability to send moisture upwards, even during dry spells. The effect is due to pressure differences between soil areas. Furthermore, tilling most certainly destroys soil horizons and hence disrupts the established flow of nutrients. A study suggests that reduced tillage preserves the crop residues on the top of the soil, allowing organic matter to be formed more easily and hence increasing the total organic carbon and nitrogen when compared to conventional tillage. The increases in organic carbon and nitrogen increase aerobic, facultative anaerobic and anaerobic bacteria populations.
- Tilling over-pumps oxygen to local soil residents, such as bacteria and fungi. As a result, the chemistry of the soil changes. Biological decomposition accelerates and the microbiota mass increases at the expense of other organic matter, adversely affecting most plants, including trees and vegetables. For plants to thrive a certain quantity of organic matter (around 5%) must be present in the soil.
- Tilling uproots all the plants in the area, turning their roots into food for bacteria and fungi. This damages their ability to aerate the soil. Living roots drill

millions of tiny holes in the soil and thus provide oxygen. They also create room for beneficial insects and annelids (the phylum of worms). Some types of roots contribute directly to soil fertility by funding a mutualistic relationship with certain kinds of bacteria (most famously the rhizobium) that can fix nitrogen.

Fukuoka advocated avoiding any change in the natural landscape. This idea differs significantly from some recent permaculture practice that focuses on permaculture design, which may involve the change in landscape. For example, Sepp Holzer, an Austrian permaculture farmer, advocates the creation of terraces on slopes to control soil erosion. Fukuoka avoided the creation of terraces in his farm, even though terraces were common in China and Japan in his time. Instead, he prevented soil erosion by simply growing trees and shrubs on slopes.

Variants of Natural Farming

Ladybirds consume aphids and are considered beneficial by natural farmers that apply biological control.

Although the term "natural farming" came into common use in the English language during the 1980s with the translation of the book *One Straw Revolution*, the natural farming mindset itself has a long history throughout the world, spanning from historical Native American practices to modern day urban farms.

Some variants, and their particularities include:

Fertility Farming

In 1951, Newman Turner advocated the practice of "fertility farming", a system featuring the use of a cover crop, no tillage, no chemical fertilizers, no pesticides, no weeding and no composting. Although Turner was a commercial farmer and did not practice random seeding of seed balls, his "fertility farming" principles share similarities with Fukuoka's system of natural farming. Turner also advocate a "natural method" of animal husbandry.

Native American

Recent research in the field of traditional ecological knowledge finds that for over one hundred centuries, Native American tribes worked the land in strikingly similar ways to today's natural farmers. Author and researcher M. Kat Anderson writes that "According to contemporary Native Americans, it is only through interaction and relationships with native plants that mutual respect is established."

Nature Farming

Japanese farmer and philosopher Mokichi Okada, conceived of a "no fertilizer" farming system in the 1930s that predated Fukuoka. Okada used the same Chinese characters as Fukuoka's "natural farming" however, they are translated into English slightly differently, as nature farming. Agriculture researcher Hu-lian Xu claims that "nature farming" is the correct literal translation of the Japanese term.

Zero Budget Farming

Zero Budget Farming is a variation on natural farming developed in, and primarily practiced in southern India.It also called spiritual farming. The method involves mulching, intercropping, and the use of several preparations which include cow dung. These preparations, generated on-site, are central to the practice, and said to promote microbe and earthworm activity in the soil. Indian agriculturist Subhash Palekar has researched and written extensively on this method.

Ecosystem-based Management

Ecosystem-based management is an environmental management approach that recognizes the full array of interactions within an ecosystem, including humans, rather than considering single issues, species, or ecosystem services in isolation. It can be applied to studies in the terrestrial and marine environments with challenges being attributed to both. In the marine realm, they are highly challenging to quantify due to highly migratory species as well as rapidly changing environmental and anthropogenic factors that can alter the habitat rather quickly. To be able to manage fisheries efficiently and effectively it has become increasingly more pertinent to understand not only the biological aspects of the species being studied, but also the environmental variables they are experiencing. Population abundance and structure, life history traits, competition with other species, where the stock is in the local food web, tidal fluctuations, salinity patterns and anthropogenic influences are among the variables that must be taken into account to fully understand the implementation of a "ecosystem-based management" approach. Interest in ecosystem-based management in the marine realm has developed more recently, in response to increasing recognition of the declining state of fisheries and ocean ecosystems. However, due to a lack of a clear definition and the diversity involved with the environment, the implementation has been lagging.

Terrestrial ecosystem-based management (often referred to as ecosystem management) came into its own during the conflicts over endangered species protection (particularly the northern spotted owl), land conservation, and water, grazing and timber rights in the western United States in the 1980s and 1990s.

Implementation

Because ecosystem-based management is applied to large, diverse areas encompassing an array of interactions between species, ecosystem components, and humans, it is often perceived as a complex process that is difficult to implement. Slocombe also noted that in addition, uncertainty is common and predictions are difficult. However, in light of significant ecosystem degradation, there is a need for a holistic approach that combines environmental knowledge and co-ordination with governing agencies to initiate, sustain and enforce habitat and species protection, and include public education and involvement. As a result, ecosystem-based management will likely be increasingly used in the future as a form of environmental management. Some suggestions for implementing ecosystem-based management and what the process may involve are as follows:

Goals and Objectives

Defining clear and concise goals for ecosystem-based management is one of the most important steps in effective ecosystem-based management implementation. Goals must move beyond science-based or science-defined objectives to include social, cultural, economic and environmental importance. Of equal importance is to make sure that the community and stake-holders are involved throughout the entire process. Slocombe also stated that a single, end-all goal cannot be the solution, but instead a combination of goals and their relationships with each other should be the focus.

As discussed by Slocombe, goals should be broadly applicable, measurable and readily observable, and ideally be collectively supported in order to be achievable. The idea is to provide direction for both thinking and action and should try to minimize managing ecosystems in a static state. Goals should also be flexible enough to incorporate a measure of uncertainty and be able to evolve as conditions and knowledge change. This may involve focusing on specific threatening processes, such as habitat loss or introduced invasive species, occurring within an ecosystem. Overall the goals should be integrative, to include the structure, organization and processes of the management of an area. Correct ecosystem-based management should be based in goals that are both "substantive", to explain the aims and importance of protecting an area, and "procedural", to explain how substantive goals will be met. some steps of ecosystem-based management may include:

Scoping

This step involves the acquisition of data and knowledge from various sources in order to provide a thorough understanding of critical ecosystem components. Sources may include literature, informal sources such as aboriginal residents, resource users, and environmental experts. Data may also be gained through statistical analyses, simulation models, or conceptual models.

Defining Indicators

Ecological indicators are useful for tracking or monitoring an ecosystem's status and can provide feedback on management progress as stressed by Slocombe. Examples may include the population size of a species or the levels of toxin present in a body of water. Social indicators may also be used such as the number or types of jobs within the environmental sector or the livelihood of specific social groups such as indigenous peoples.

Setting Thresholds

Tallis et al. suggest setting thresholds for each indicator and setting targets that would represent a desired level of health for the ecosystem. Examples may include species composition within an ecosystem or the state of habitat conditions based on local observations or stakeholder interviews. Thresholds can be used to help guide management, particularly for a species by looking at the conservation status criteria established by either state or federal agencies and using models such as the minimum viable population size.

Risk Analysis

A range of threats and disturbances, both natural and human, often can affect indicators. Risk is defined as the sensitivity of an indicator to an ecological disturbance. Several models can be used to assess risk such as population viability analysis.

Monitoring

Evaluating the effectiveness of the implemented management strategies is very important in determining how management actions are affecting the ecosystem indicators. Evaluation: This final step involves monitoring and assessing data to see how well the management strategies chosen are performing relative to the initial objectives stated. The use of simulation models or multi-stakeholder groups can help to assess management.

It is important to note that many of these steps for implementing ecosystem-based management are limited by the governance in place for a region, the data available for assessing ecosystem status and reflecting on the changes occurring, and the time frame in which to operate.

Challenges

Because ecosystems differ greatly and express varying degrees of vulnerability, it is difficult to apply a functional framework that can be universally applied. These outlined steps or components of ecosystem-based management can, for the most part, be applied to multiple situations and are only suggestions for improving or guiding the challenges involved with managing complex issues. Because of the greater amount of influences, impacts, and interactions to account for, problems, obstacles and criticism often arise within ecosystem-based management. There is also a need for more data, spatially and temporally to help management make sound decisions for the sustainability of the stock being studied.

The first commonly defined challenge is the need for meaningful and appropriate management units. Slocombe noted that these units must be broad and contain value for people in and outside of the protected area. For example, Aberley suggests the use of "bioregions" as management units, which can allow peoples involvement with that region to come through. To define management units as inclusive regions rather that exclusive ecological zones would prevent further limitations created by narrow or restricting political and economic policy created from the units. Slocombe suggests that better management units should be flexible and build from existing units and that the biggest challenge is creating truly effect units for managers to compare against.

Another issue is in the creation of administrative bodies. They should operate as the essence of ecosystem-based management, working together towards mutually agreed upon goals. Gaps in administration or research, competing objectives or priorities between management agencies and governments due to overlapping jurisdictions, or obscure goals such as sustainability, ecosystem integrity, or biodiversity can often result in fragmented or weak management. In addition, Tallis stated that limited knowledge of ecosystem components and function and time constraints that can often limit objectives to only those that can be addressed in the short-term.

The most challenging issue facing ecosystem-based management is that there exists little knowledge about the system and its effectiveness. Slocombe stated that with limited resources available on how to implement the system it is hard to find support for its use.

Slocombe said that criticism of ecosystem-based management include its reliance on analogy and comparisons, too broadly applied frameworks, its overlap with or duplication of other methods such as ecosystem management, environmental management, or integrated ecosystem assessment, its vagueness in concepts and application, and its tendency to ignore historical, evolutionary or individual factors that may heavily influence ecosystem functioning.

Tallis stated that ecosystem-based management is seen as a critical planning and management framework for conserving or restoring ecosystems though it is still not widely implemented. An ecosystem approach addresses many relationships across spatial, biological, and organizational scales and is a goal-driven approach to restoring and sustaining ecosystems and functions. In addition, ecosystem-based management involves community influence as well as planning and management from local, regional and national government bodies and management agencies. All must be in collaboration in order to develop a desired future of ecosystem conditions, particularly where ecosystems have undergone radical degradation and change. Slocombe said that to move forward, ecosystem-based management should be approached through adaptive management, allowing flexibility and inclusiveness to deal with constant environmental, societal, and political change.

Marine Systems

Ecosystem-based management for marine environments moves away from the traditional strategies in which single species and single sectors are managed individually; rather it is an integrated approach which considers all key activities, particularly anthropogenic, that affect marine environments. The management must take into account the life history of the fish being studied, its association with the surrounding environment, its place in the food web, where it prefers to reside in the water column and how it is affected by human pressures. The objective is to ensure sustainable ecosystems, thus protecting the resources and services they provide for future generations.

In recent years there has been increasing recognition of disruption to marine ecosystems resulting from climate change, overfishing, nutrient and chemical pollution from land runoff, coastal development, bycatch, habitat destruction and other anthropogenic sources. There are very clear links between human activities and marine ecosystem functioning; this has become an issue of high importance because there are many services provided by marine ecosystems that are declining as a result of these impacts. These services include the provision of food, fuel, mineral resources, pharmaceuticals, as well as opportunities for recreation, trade, research and education.

Guerry has identified an urgent need to improve the management of these declining ecosystems, particularly in coastal areas, to ensure a sustainable future. Human communities depend on marine ecosystems for important resources, but without holistic management these ecosystems are likely to collapse. It has been suggested that the degradation of marine ecosystems is largely the result of poor governance and that new approaches to management are required. The Pew Oceans Commission and the United States Commission on Ocean Policy have indicated the importance of moving from current piecemeal management to a more integrated ecosystem-based approach.

Stock Assessment

In marine environments, it can be increasingly difficult to determine what part of the population a certain stock comes from. In the case of the Southern bluefin tuna, it is all one taxonomic stock while in the Pacific Northwest, each river has its own stock of pacific salmon and thus, the clear definition of a particular stock can be quite challenging. With knowledge on the specific species life history, stock assessment can be much easier and therefore, knowledge on the animals breeding patterns, juvenile state and morphological patterns are essential.

Dead salmon in spawning season.

Bottom Up or Top Down

When trying to analyze the specific species, the manager needs to fully understand where the species is in terms of the food web. In the North Sea, the puffin population crashed and no one was quite sure why. After further analysis, they realized that the decline in seabird numbers was due to the sandeel population being severely depleted due to fishing pressures and this is an example of top down. In Alaska, the sea otter was hunted mainly for its pelt and was nearly pushed to extinction. As their numbers dwindled, the kelp beds were disappearing as well at an alarming rate due to the sea urchins having no predators and eating all of the kelp. As the sea otters were removed, the rest of the ecosystem in the food chain beneath them were affected, an example of top-down.

Post-dip pose.

Bycatch

In the Gulf of Mexico, the red snapper is one of the most important economical species there is. However, it is a very difficult species to manage due to the bycatch associated with the shrimping industry. The juveniles are caught in the smaller mesh size used for the shrimp and thus, a large portion of the red snapper population is subject due to mortality not associated with the red snapper fishery.

Key Elements

Connections

At its core, ecosystem-based management is about acknowledging interdependency connections, including the linkages between marine ecosystems and human societies, economies and institutional systems, as well as those among various species within an ecosystem and among ocean places that are linked by the movement of species, materials, and ocean currents. Of particular importance is how these factors all react and involve each other. In the Caribbean, the spiny lobster is managed based on a classic population model that for most fishery species works quite well. However, this species will grow and then halt its growth when it need to molt its shell and thus instead of a continuous growth cycle, it will pause its growth and invest its energy in a new shell. To further complicate matters, it slows this process down as it gets older to invest more energy into reproduction thus further deviating itself from the von Bertalanffy model of growth that was applied to it. The more information we can gather about an ecosystem and all of the interconnected factors which affect it, the more capable we will be of better managing that system.

Cumulative Impacts

Ecosystem-based management focuses on how individual actions affect the ecosystem services that flow from coupled socio-ecological systems in an integrated fashion, rather than considering these impacts in a piecemeal manner. Loss of biodiversity in marine ecosystems is an example of how cumulative effects from different sectors can impact on an ecosystem in a compounding way. Overfishing, coastal development, filling and dredging, mining and other human activities all contribute to the loss of biodiversity and therefore degradation of the ecosystem. Work is needed prior to the carrying out of the research to understand the total effects that each species can have on each other and also on the environment. It must be carried out every year as well as species are changing their life history traits and their relationship with the environment as humans are continually modifying the environment.

Interactions between Sectors

The only way to deal with the cumulative effects of human influences on marine ecosystems is for various contributing sectors to set common goals for the protection or management of ecosystems. While some policies may only affect a single sector, others may affect multiple sectors. A policy for the protection of endangered marine species, for example, could affect recreational and commercial fisheries, mining, shipping and tourism sectors to name a few. More effective ecosystem management would result from the collective adoption of policies by all sectors, rather than each sector creating their own isolated policies. For example, in the Gulf of Mexico there are oil rigs, recreational fisheries, commercial fisheries and multiple tourist attractions. One of the main fisheries is that of the Red Snapper which inhabits much of the Gulf and employs thousands of people in the commercial and recreational fishery. During the Deepwater oil spill it became abundantly obvious that it negatively affected the population numbers as well as the integrity of the catch that was being made. The species not only suffered higher mortality rates but the market was less trusting of the product. An environmental disaster interacted with the commercial, recreational, and economic sector for a specific species.

Changing Public Perceptions

Not all members of the public will be properly informed, or be fully aware, of current threats to marine ecosystems and it is therefore important to change public perceptions by informing people about these issues. It is important to consider the interest of the public when making decisions about ocean management and not just those who have a material interest because community support is needed by management agencies in order to make decisions. The Great Barrier Reef Marine Park Authority (GBRMPA) faced the issue of poor public awareness in their proposed management strategy which included no-take fishing zones. Olssen addressed this problem by starting a 'reef under pressure' information campaign to prove to the public that the Great Barrier Reef is under threat from human disturbances, and in doing so were successful in gaining public support.

Bridging Science and Policy

To ensure that all key players are on the same page, it is important to have communication between managers, resource users, scientists, government bodies and other stakeholders. Leslie and McLeod stated that proper engagement between these groups will enable the development of management initiatives that are realistic and enforceable as well as effective for ecosystem management. If certain small-scale players are not involved or informed, it is highly unlikely and equally challenging to get them to cooperate as well as to follow the rules that need to be put in place. It is of the utmost importance to have every stakeholder involved with every step of the process to increase the cohesion of the process.

Embracing Change

Coupled social-ecological systems are constantly changing in ways that cannot be fully predicted or controlled. Understanding the resilience of ecosystems, i.e. the extent to which they can maintain structure, function, and identity in the face of disturbance, can enable better prediction of how ecosystems will respond to both natural and anthropogenic perturbations, and to changes in environmental management. With how much modification humans are doing to environments, it is important to understand these changes on a yearly basis as well. Some species are changing their life histories, Flounder, due to the increased pressures that humans are placing on the environment. Thus, when a manager or government does an assessment on the ecosystem for a given year, the relationship that a species has to others can change very quickly and thus negate the model that you use for an ecosystem very quickly if not redefined.

Multiple Objectives

Ecosystem-based management focuses on the diverse benefits provided by marine systems, rather than on single ecosystem services. Such benefits or services include vibrant commercial and recreational fisheries, biodiversity conservation, renewable energy from wind or waves and coastal protection. The goal is to provide a sustainable fisheries while incorporating the impacts of other aspects on that resource. When managed correctly, an ecosystem-based model can greatly improve not only the resource being managed, but those associated with it.

Learning and Adaptation

Because of the lack of control and predictability of coupled social-ecological systems, an adaptive management approach is recommended. There can be multiple different factors that must be overcome (fisheries, pollution, borders, multiple agencies, etc.) to create a positive outcome. Managers must be able to react and adapt as to limit the variance associated with the outcome.

Great Bear Rainforest - Canada

The Land and Resource Management Planning (LRMP) was implemented by the

British Columbia Government (Canada) in the mid-1990s in the Great Bear Rainforest in order to establish a multiparty land-use planning system. The aim was to "maintain the ecological integrity of terrestrial, marine and freshwater ecosystems and achieve high levels of human well-being". The steps described in the programme included: protect old-growth forests, maintain forest structure at the stand level, protect threatened and endangered species and ecosystems, protect wetlands and apply adaptive management. MacKinnon highlighted that the main limitation of this program was the social and economic aspects related to the lack of orientation to improve human well-being.

The Great Lakes - Canada and United States

A Remedial Action Plan (RAP) was created during the Great Lakes Water Quality Agreement that implemented ecosystem-based management. The transition, according to the authors, from "a narrow to a broader approach " was not easy because it required the cooperation of both the Canadian and American governments. This meant different cultural, political and regulatory perspectives were involved with regards to the lakes. Hartig et al. described eight principles required to make the implementation of ecosystem-based management efficacious: "broad-based stakeholder involvement; commitment of top leaders; agreement on information needs and interpretation; action planning within a strategic framework; human resource development; results and indicators to measure progress; systematic review and feedback; and stakeholder satisfaction".

Scallop Aquaculture in Sechura Bay and Peru

Peruvian Bay Scallop is grown in the benthic environment. Intensity of the fishery has caused concern over recent years and there has been a shift to more of an environmental management scheme. They are now using food web models to assess the current situation and to calibrate the stocking levels that are needed. The impacts of the scallops on the ecosystem and on other species are now being taken into account as to limit phytoplankton blooms, overstocking, diseases and overconsumption in a given year. This study is proposed to help guide both fisherman and managers in their goal of providing long term success for the fishery as well as the ecosystem they are utilizing.

Dam Removal in the Pacific Northwest

The Elwha dam removal in Washington state is the largest dam removal project in the United States. Not only was it blocking several species of salmon from reaching their natural habitat, it also had millions of tons of sediment built up behind it.

Sustainable Ecosystem Management

Sustainable Ecosystem Management (SEM) is one of the 'four challenges' for ecosystem services (ES) to be addressed by Open NESS; the others relate to human well-being, competitiveness and governance. The challenge is to manage ecosystems in such a way that they can sustainably deliver an optimal combination of ecosystem services both today and into the future, with resilience to environmental and social change.

The term 'ecosystem' was introduced in 1935, to describe the entire system of living organisms and interacting abiotic factors such as air, water and minerals occupying a given space. Prior to this, land management had typically been geared towards either the conservation of single iconic species, or the production of single resources such as timber or food. However, increasing recognition of the way in which species interacted with each other and with the surrounding environment led to the concept of ecosystem management, which originated with four principles established by the Ecological Society of America: protect entire habitats as well as particular species; maintain a wide range of native ecosystem types within each region; manage for resilience to disturbances; and establish buffer zones around core reserves. These ideas rose to prominence during the 1970s and 80s, triggered by debate over the future of species such as the grizzly bear. This led to wider recognition of the need to provide a large enough area of primary habitat to sustain the largest carnivore in a region, even if that area crosses administrative boundaries (such as national park or even national boundaries), thus requiring increased co-operation and co-ordination between agencies. There was also growing awareness that resource extraction was leading to accelerating biodiversity loss and ecosystem degradation, but this was coupled with increasing acceptance that conservation had to take account of human social and economic needs.

By the late 1980s, the term 'Ecosystem Management' was widely used to describe an approach that tried to tackle all these issues. There are many competing definitions, but one of the most widely accepted is:

> Ecosystem management is the application of ecological science to resource management to promote long-term sustainability of ecosystems and the delivery of essential ecosystem goods and services to society.

There is no separate definition of 'sustainable ecosystem management', but the concept of sustainability is at the heart of the definition above, with its emphasis on balancing human needs with the long term sustainability of ecosystems. This is consistent with the well-known definition of sustainability as 'meeting the needs of the present and local population without compromising the ability of future generations or populations in other locations to meet their needs'. Over the last two decades, this has been reinforced by the concept of 'socio-ecological systems', which recognises that humans and their social and political systems are an integral part of ecosystems, and that successful ecosystem management therefore requires sustainable and equitable governance. The key characteristics of Ecosystem Management include:

- It considers entire ecosystems, with their complex connections, rather than individual sites or species.

- It aims to balance human social, economic and cultural needs (the delivery of essential ecosystem services) with ecosystem sustainability (maintaining healthy, productive and resilient ecosystems in the long term).
- It involves all stakeholders in collaborative decision-making, including scientists, national and local government, ngos, business, local residents and the public.
- Conflicts will arise due to different stakeholder needs: these must be resolved through negotiation.
- It is an interdisciplinary approach bringing together science, resource management, planning, economics, sociology, law and politics.
- It involves co-operation between different agencies, crossing political and organisational boundaries.
- It is place-based, but also considers wider objectives such as cumulative environmental, social or economic effects at regional, national or global scale.
- It uses a flexible adaptive management approach, with continual monitoring, learning, feedback and adjustment of goals and strategies to meet changing needs and incorporate new information.

Procedural sustainability is a key feature of sustainable ecosystem management. Co-operation between stakeholder groups allows socio-ecological systems to innovate and adapt to change. Goals should be set as a result of negotiation between all stakeholders, and indicators should be chosen carefully to match the goals. Adaptive management is the key to dealing with the highly complex, uncertain and unpredictable nature of socio-ecological systems. Typical steps in setting up a sustainable ecosystem management system are listed below.

- Define the boundaries of the ecosystem to be managed.
- Assess the ecosystem, using all available sources including local and traditional knowledge. Determine key components and their interactions (species, habitats, abiotic factors); assess the condition and integrity of the ecosystem; evaluate supply and demand of ecosystem services; and identify carrying capacity, potential thresholds and tipping points.
- Identify stakeholders, understand their use of different ES and the values they attach to those ES, and define management goals through a collaborative/negotiation process involving all stakeholders.
- Select suitable indicators to measure achievement of the goals.

- Design and implement actions to achieve the goals, e.g. defining protected areas and buffer zones, setting appropriate levels of resource extraction in certain areas (typically excluding high-integrity areas), restoring degraded ecosystems, and improving connectivity.
- Monitor results, perhaps using GIS and remote sensing to help assess impacts on land cover.
- Regularly review and adjust the strategy, in consultation with stakeholders, in the light of environmental, social and political developments and new scientific information.

Indicators for Sustainable Ecosystem Management

Selection of suitable indicators for monitoring and managing ecosystems is crucial, in order to ensure that different needs are balanced and ecosystems are not being over-exploited. The choice of indicators will vary depending on the goals prioritised by the stakeholders, but they can reflect the different stages of the ES cascade model: ecosystem structure (e.g. species diversity), function (e.g. water and nutrient cycling) and ES delivery (e.g. visitor numbers, water quality regulation), as well as socio-economic indicators such as jobs or incomes.

To assess long term sustainability of ecosystems, and detect signs of over-exploitation, indicators of ecosystem health and resilience are needed. Rapport et al.define ecosystem health as being based on vigour (e.g. productivity), organisation (e.g. functional diversity) and resilience (the ability to maintain structure and function under stress). As more 'natural' systems tend to be more diverse and more able to cope with change, resilience is related to ecosystem integrity, defined as the degree to which ecosystems have changed from their natural state due to human intervention. However, the stability of ecosystem functions is now viewed as being more important than 'naturalness' per se. For example, Palmer and Febria recommend monitoring ecosystem structure and function to determine ecosystem health and resilience, while Kandziora et al. adopt a definition of integrity that is based on structure and function, not on naturalness.

'Ecosystem Management' is synonymous with a number of other terms, including 'ecosystem-based management' (EBM), 'ecological stewardship' and the 'ecosystem approach'. It is distinct from 'natural resource management', which tends to refer to management of a particular resource (e.g. timber, water) for human use, but it is similar to the more holistic 'integrated natural resource management'. Sustainable Forest Management (SFM) is an evolution of this approach that is similar to Ecosystem Management, though Sayer and Maginnis point out subtle differences linked to the fact that SFM is typically led by foresters whose main aim is to maximise sustainable yield, whereas Ecosystem Management is typically led by people seeking to protect ecosystems. There is some overlap of 'Ecosystem Management' with other approaches that consider ecosystems at a wider spatial scale, e.g. 'integrated river basin management', or 'landscape level conservation' (e.g. using wildlife corridors to improve connectivity). There are also similarities with terms that link human use of ecosystems with environmental protection, e.g. 'sustainable development' or 'integrated conservation and development'.

Controversy over the Definition and Main Goal

Although the definition of Ecosystem Management presented above clearly emphasises the need to balance human needs and long term ecosystem sustainability, the concept has attracted controversy. A review by Lackey concluded that the definition and aims adopted by different groups tends to reflect their personal views. Conservationists tend to see ecosystem management as a powerful tool to protect ecosystems from damage by humans, whereas many natural resource managers see it as justifying the further exploitation of ecosystems to provide goods and services for humans. Lackey cites one as saying:

> The debate between 'biocentric' conservationists, who emphasise the intrinsic value of nature, and 'anthropocentric' conservationists who argue that the primary function of nature is to provide services for humans, is still ongoing today. Similarly, while many conservationists emphasise the importance of integrity in providing resilience to change, resource managers tend to argue that 'preservation from all change is not usually the main goal'. However, the process of ecosystem management described above is explicitly designed to arrive at a sensible balance between these competing viewpoints, through a process of negotiation and mutual learning.

One approach to reconciling the use of ecosystems by nature and by people is that adopted by Brussard et al., who propose different management goals for 'high-integrity ecosystems', 'impacted ecosystems' and 'cultural ecosystems', as follows.

High integrity ecosystems are those which are relatively unchanged by humans. They are rich in native species and tend to be well-adapted to the prevailing patterns of disturbance. They are also an important source of rare species which may become more important for maintaining biological functions and delivering ecosystem services when environmental conditions change, and they provide baselines for assessing the relative condition or state of other ecosystems. These ecosystems – including national parks, nature reserves, wilderness areas, critical watersheds and multiple-use lands of particularly high conservation value - should be maintained in, or restored to, high-integrity states.

Impacted ecosystems are natural or semi-natural areas that have lost much of their original integrity due to human activity, but they still provide many services. These lands should be managed for ecosystem viability or 'health'. Impacted ecosystems may be considered viable if they meet the following criteria:

- Current utility: the ecosystem is providing the goods or services expected from it with reasonable efficiency (e.g., preventing erosion, detoxifying wastes);
- Future potential: present uses are not disrupting processes that generate and maintain the desired composition, structure, and functional organization of the ecosystem;
- Containment: present uses or current conditions do not degrade areas beyond the system's borders;
- Resilience: the system has the capacity for self-maintenance and self-regeneration i.e., it can maintain the desired structure, composition, and functional organization after moderate external stresses or perturbations.

For example, an impacted stream ecosystem might include non-native fish species, but it would still be considered viable if: it is providing the services expected of it (e.g. fresh water, fishing); the remaining native species in the system are stable; it is not contributing to the degradation of other streams (e.g. by exporting excess nutrients or non-native species); and it has resilience (i.e. it can self-organise and adapt in the face of change, e.g. continuing to provide the expected services after a serious flood or drought).

Cultural ecosystems, such as agricultural fields and urban areas, have been transformed completely by human activity. They are important to human wellbeing and livelihoods, and for delivery of certain provisioning ecosystem services such as food, but require constant input of energy and materials. These ecosystems are not self-sustaining and probably not viable: they have current utility for humans, but no resilience and little future ecological potential; most also fail the containment criterion. Brussard et al. do not specify what approach should be applied to these areas, but it seems likely that they would present considerable opportunities for ecosystem enhancement, e.g. through tackling pollution or adopting 'nature based solutions'.

Current Status

Over the last two decades, the concept of ecosystem management has been taken up in a range of situations, though there still does not appear to be a universally accepted definition or set of guidelines for implementation. Sustainable Forest Management (SFM) is now widely practised, building on the Forest Principles that were adopted at the United Nations Conference on Environment and Development in Rio de Janeiro in 1992. Ecosystem Management can be seen as a logical progression from SFM: it is applied to a wider range of goods and services (as it is not restricted to forests), but it is still an operational approach linked to a set of tools and methodologies. It was adopted by the US Forest Service in the late 1980s, generating a large body of literature in the 1990s, and is enshrined in the 2008 Planning Rule. Ecosystem-based management has also been applied to coasts and oceans including in the 2008 EU Marine Strategy Framework Directive. More recently, the concept was incorporated into the EU Biodiversity Strategy (sustainable management of agriculture is mentioned in Target 3a; SFM in Target 3b) and SFM is mentioned in the EU Forest Strategy. At a broader level, twelve general principles and five points of operational guidance for the 'Ecosystem Approach' have been developed by the Convention on Biological Diversity (CBD), which has also provided a sourcebook of guidelines for applying the approach, and a database of case studies. SEM is related to the CBD's Aichi Targets, especially Strategic Goals C and D. Concepts and approaches to SEM will be key in the delivery of the post-2015 SDGs adopted by the UN General Assembly and accepted by the EU and all member states (ecosystem integrity is an over-arching principle for all 17 SDGs).

GIS and Natural Resource Management

The economic, social and cultural viability of any nation is mainly determined by the land and water resources that it has. These natural resources are essential to the economy of a nation since they play a critical role in the provision of employment, they are a source of raw materials for various industries, acts as a source of food and income, medicine as well as energy. The aesthetic beauty that relates to natural resources is always regarded as the cultural representation of nations. For instance, when one talks about the grand canyon, the picture of the United States comes into mind. When one talks about oil wells, one thinks about the Middle East. Consequently, when one talks about the big 5 game animals, the picture of Africa, especially the Maasai Mara in Kenya comes into play.

However, it is essential for nations to learn how to use these resources in a sustainable manner to ensure that their benefits are enjoyed in the present as well as future generations. This is because, these resources can be depleted if they are not utilized in an effective and efficient manner. At the present moment, the utilization of the resources present in the world has been overstretched due to the ever rising population of human beings. It is as a result of this population pressure that forest cover all over the world has declined due to human encroachment. This has greatly increased human/wildlife conflicts and encouraged the development of desert like conditions. Ultimately, the depletion of natural resources has led to an increase in the cost of living, changes in weather patterns and a decline in the economic, social and cultural benefits that were accrued as a result of their utilization.

Due to these constraints, there is a dire need to ensure that these resources are effectively managed. A lot of management practices have been advanced in the field of natural resource managed to achieve this goal. Some of them have been effective while others have failed to achieve the desired outcome. However, with the current trend in the advancement in the field of information technology, natural resource managers have now laid a lot of emphasis on the use of remote sensing and GIS technologies in the management of natural resources. These technologies provide a platform through which managers can generate informative data and information that can be used to make sound decisions for sustainable development. For over three decades now, the use of land-based satellite remote sensing has been in application in almost all realms of earth sciences. All through this time, remote sensing and GIS technologies have always generated informative data that have led to the sustainable management.

Use of Remote Sensing

In natural resource management, remote sensing and GIS is mainly used in the mapping process. These technologies can be used to develop a variety of maps. Examples include:

- Land cover maps.
- Vegetation maps.
- Soil maps.
- Geology maps.

However, before these maps are developed, there are a variety of data that need to be collected and analysed. Most of this data is collected with the help of remote sensing technology. Data can be collected using either ground photographs, aerial photographs or satellite photographs of the area of study. The choice of the photograph usually depends on the topography of the area of study and the aim of the study. For instance, aerial photographs (vertical or oblique) are always useful when spatial data need to be collected in the same area of study within intervals (hours, days, seasons, years etc). This form of data collection shows the variations of the area of study within different periods of time. Satellite photographs can also be used to collect relevant data for the study. These types of photographs are however superior to aerial photographs in the sense that they have higher spectral, spatial, radiometric and temporal resolutions. Thus, satellite images are more detailed hence a lot of data can be generated from them. However, for remote sensing data to be effective, it needs to be incorporated together with topographical maps that show the variation of climate, soils, and other factors.

The visual and digital data that has been collected is usually analysed to generate a pre-field map. Various components and elements of the data is analysed. According to Elias (n.d.), elements such as tone, texture, pattern, association, size and shape are essential in the analysis process. These elements bring about a detailed view of the area of study. The pre-field map that has been generated together with the results from the analysis of the various elements is used to determine the characteristics of different elements and themes found on the ground.

Ground verification of the collected data is a critical process. To ensure that it is carried out in the most effective and efficient manner, the study area is usually divided into quadrants or transects. This is done to ensure that the interpreted elements of the satellite data conform to the ground characteristics. The data that is collected on the ground is geo-referenced with the help of a GPS to ensure that its corresponding location can be accurately identified in the images that were collected earlier. Additionally, field points identified in the images are visited to verify information regarding the state of the vegetation, geomorphology, topography, soils, and so on. With the use of the pre-field map that was generated, information from ground verification procedure and any other secondary source that might have been used, the final map is usually prepared. The scale of the map is also variable depending on the nature and extent of the study and the goals that it aims to achieve.

Application of GIS Data

Forest Management

Over the last century, the forest cover of the world has declined at an alarming rate. Being a renewable resource, forest cover can be regenerated through sustainable management. Hence, with the help of remote sensing and GIS data, a forest manager can generate information with regards to forest cover, types of forest present within the area of the study, human encroachment into forest land/protected areas, encroachment of desert like conditions and so on. This information is critical in the development of forest management plans and in the process of decision making to ensure that effective policies have been put in place to control and govern the manner in which forest resources is utilized.

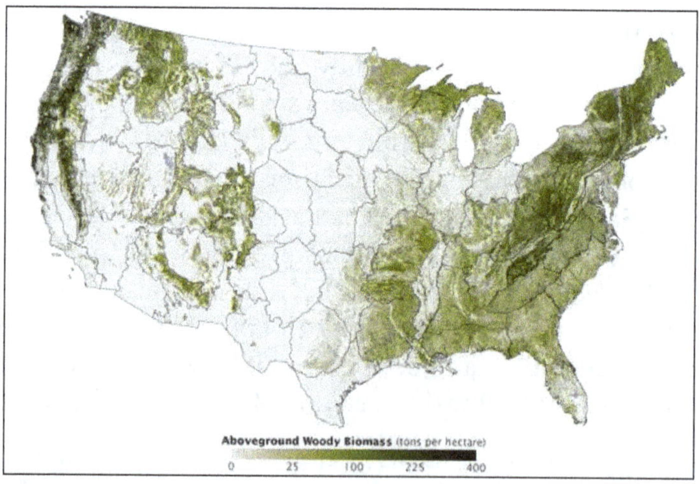

Aboveground woody biomass data by josef kellndorfer and wayne walker.

Watershed Management

Water as a resource has been diminishing over the years. In Africa and other developing nations, the availability of clean water has been always scarce. Water management has therefore been a challenge in developing nations. However, with the use of satellite data, water bodies such as rivers, lakes, dams and reservoirs can be mapped in 3D with the help of GIS technology. This data can be used in the sustainable management of water bodies since respective authorities can decide which regions need effective protection and management. At the same time, decisions regarding the most effective means of utilization of these regions can always be arrived at.

Combat Desertification

Geospatial data can be used to determine the soil types present in a given area and nutrient availability. Negative change can always be identified once this data is collected over a long period of time. GIS data can also be used to determine the land use practices within a given area and vegetation constitution and the impact that they have on the environment. Consequently, slope information of a region can also be determined with the use of GIS data. With all this information, an individual can easily determine whether desert like conditions are encroaching in an area. If desert like conditions have been identified, its impacts and intensity shall be analysed in order to decide on whether artificial or natural methods shall be used to combat the situation.

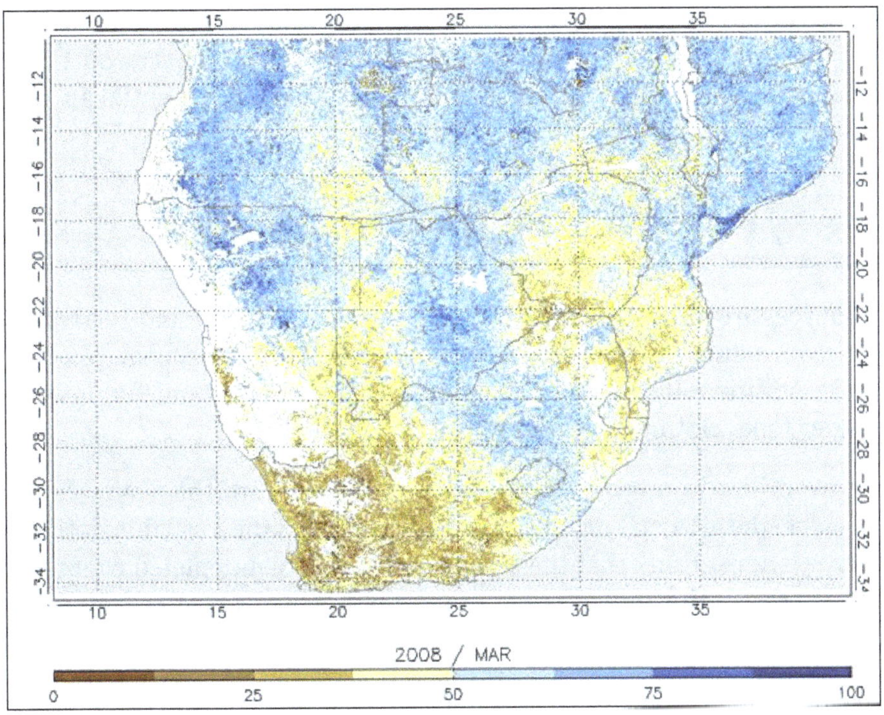

Surface soil moisture at 1 km resolution retrieved from the advanced synthetic aperture radar (asar) instrument on esa's envisat satellite, operating in global mode, for the southern african development community region. These results were obtained from esa's share project.

Biodiversity Management

Geospatial data can also be used in the management of flora and fauna within protected areas. Ground and aerial photographs, for instance, are essential in this practice.

Aerial and satellite photographs can be used to determine the presence and distribution of vegetation within a protected area. These photos can also be used to determine the presence and distribution of invasive species within an ecosystem. This information is essential as it determines the amount of cover and food that is present, particularly for herbivores during various seasons of the year. Aerial photographs can be used to ease the process of counting during animal census activities. The stop capability of photographs eases this process. It is always essential for protected area managers to determine the population and distribution of various animal species within a protected area to ensure that they have enough food and water, to eliminate the chances of overstocking that might lead to soil erosion and to ensure that a balance within the ecosystem is arrived at. Geospatial data can also be used to show human encroach-ment into protected areas as well as animal activities outside protected areas. This data critical especially in the process of resolving human/wildlife conflicts. Finally, the use of GPS technology can be applied to monitor the movement of endangered species as well as newly introduced species to determine their progress as well as protecting them from poachers. Finally, geospatial data can be used to carry out environmental impact assessment (EIA) of various projects carried out within protected areas. Projects such as building of roads, buildings, pipe ways, dams, and so on might have various effects on the flora and fauna of the ecosystem. Thus, geospatial data has become essential in biodiversity management.

Ways to Conserve Natural Resources

Natural resources are utilized by human beings either directly or indirectly for survival. The examples of natural resources are air, soil, water, sunshine, coal, plants, animals, and minerals. Additionally, nature acts as the sole provider for our fundamental needs which includes food, clothing and shelter.

Everything we utilize in our day to day lives is drawn from the available natural resources. Besides, the artificial resources are derived from the available natural resources. However, we never think that these resources might get depleted eventually if used irresponsibly. The surging global population, rapid urbanization and industrialization have created an increased demand on the available resources.

Proper steps need to be taken to ensure that we continue enjoying these resources and failure to do that, the future generation will face tremendous hardships. Natural resource conservation ensures that the ecological balance is maintained.

Use of alternative sources of power such as solar and wind energy: These alternative sources of energy are bio friendly particular because they do not produce harmful gases that damage the ozone layer: They are better compared to burning fossils such as coal and charcoal. They are also cheap to use, not easily depleted and are renewable.

Plant trees to prevent soil erosion: This entails planting trees and vegetation to control soil erosion caused by wind and water. Trees and vegetation are essential in the maintenance of the ecosystem. They also act as home for most insects, birds and some symbiotic plants. This creates a habitat for wildlife therefore conserving wildlife altogether.

Practicing of judicious ways to conserve water in our homes: This entails simple practices like ensuring that taps are closed when they are not in use. Taking less time in the shower aids to conserve gallons of water per month. Reusing some of the water for watering the kitchen gardens in our homes is also an important practice.

Use pipelines to transport oil: Many oil leaks are as a result of transporting oil using ships. The oil leaks are detrimental to the flora and fauna. Most of this species die as a result of this oil spills. Pipelines' are examples of modern ways of transporting oil.

Growing of vegetation in catchment areas: Catchment areas act as a source of water that flows in the streams, rivers and oceans. Vegetation in the catchment areas allows for sufficient percolation of water into the deep soil layers allowing for the formation of ground water. Water is life. Protecting the catchment areas ensures that our taps do not run dry.

Treatment of industrial wastes and sewages before they are released in the water bodies: Rapid industrialization has resulted to wastes that are harmful to the ecosystem. The release of these effluents directly into the water bodies has led to massive water pollution to some areas in the world especially in developing countries. Therefore, industrial and human waste should be treated to prevent thermal and chemical pollution of water.

Rain harvesting: This is an important practice of ensuring water conservation. It is done by harvesting of rainwater during the wetter seasons of the year and using it during the dry seasons. This minimizes the usage of water from the water-bodies. The water can be used for irrigation in our farms and also watering the lawns in our homes.

Practice of in-situ or on site conservation of wildlife: This entails conservation of animals and plants in their natural habitats. It is also involves protecting the protected areas set aside. They include areas such as wildlife sanctuaries, parks, biosphere reserves among others. Certainly, this will help conserve the endangered plant and animal species.

Practice Ex-situ or offsite conservation of wildlife: It involves the conservation of animals and plants outside the natural habitats. These includes areas such as pollen banks, DNA banks, zoos, seed banks, botanical gardens, tissue culture banks among others.

Formulation of policies to regulations to curb poaching: Poachers continue to kill many animals such as elephants, rhinos, leopards for their horns and skin which are sold off in the black market: Poachers are a major threat to our biodiversity as they are slowly making some species extinct. These regulations will ensure that poaching is done away with.

Practice judicious ways of conservation energy: Such practices include switching off the lights when not in use, unplugging electrical appliances when not in use. Plugged-in appliances continue to use electricity even when not in use. Other practices include spending less time when taking hot showers.

Use of biogas in our homes: Around the World, Liquefied Petroleum Gas (LPG) is the most rampant source of fuel in our homes today. Continued LPG use results to the depletion of the oil reserves, biogas is therefore an alternative. Biogas is mainly produced from cattle dung, biogas plants are a source of both biogas and manure.

Use of bio-fuels: For more than a century, fossil fuels have been a major source of energy. However, they are depleting rapidly, this calls for alternative sources of fuel such as bio-fuels which are mainly from plant species. Bio-fuels are known to be bio friendly and they reduce the occurrence of air pollution.

Ensure the recycling of wastes: These wastes include plastics, paper bags that have resulted to tones of garbage. Recycling entails re-manufacturing of already used materials. This reduces the amount of waste available reducing soil and water pollution.

Planting trees in home compounds: Trees provide shade and perfect air conditioning for our homes. Trees ensure that there is clean air for us to breathe creating a perfect air conditioner. This helps to cut down the electricity costs from the use of air conditioning appliances.

Make use of electronic mails: Electronic mails are paperless and present a good way to minimize the usage of paper. Technology has made this possible reducing the usage of paper and envelops. This has reduced the production of paper and also minimized cutting down of trees.

Purchase hybrid cars instead of the conventional cars: Hybrid cars use a combination of electricity and minimal amounts of gas to run them. This is a break from the use of petroleum products that are mostly used to run cars. The hybrid cars are a great deal in conserving fossil fuels.

Use earth-bags instead of plastic and paper bags: Most of the waste comes from excessive use of plastic bags. Carry an earth-bag each time you visit the grocery store. Products should also be wrapped in materials that can be recycled.

Use energy saving fluorescent bulbs: Energy saving bulbs may cost more money than the old bulbs. However, they will minimize electricity bills thus saving energy. They also last longer ensuring that fewer purchases are made.

Industries can ensure the make production efficient to reduce wastage: This will ensure limited harvesting of natural resources and also reduce wastage.

Use ceramic cups to consume the daily cup of coffee rather than using disposable mugs: This ensures that there is less waste and also ensures the overall conservation of the ecosystem.

Water the lawns and farms in the evening: Watering the farm when it is dry and hot results to increased water evaporation and a lot of water is used for the same. During the evening, the weather is much cooler reducing evaporation thus conserving water.

Reuse old furniture: It is common to dispose of old furniture and opt for new furniture. The old furniture should be sold off for use or donated to charity where they can be reused. The old furniture can also be re-sculptured and redecorated to save wood. This will reduce deforestation.

Practice crop rotation: Planting the same crops for a long period of time reduces soil fertility. The practice of crop rotation will restore and maintain soil fertility thus conserving the soil.

Translocation of wild animals: The growing population has led to human encroaching on the wildlife habitat. This has resulted to human-animal conflict where the wildlife are killed by humans as a way of protecting themselves from them. Translocation involves moving wild animals to adjacent areas and fencing to curb the conflict.

Encourage the use of drip irrigation: Some of the water from daily activities may be used in grip irrigation. This promotes water conservation.

Establish special schemes to preserve endangered plant and animal species: This includes botanical gardens, sanctuaries that may be established to protect the endangered species so that they can be available for future generations.

Constructions of reservoirs: This will regulate the amount of water that is used daily. The dams also act as a source of hydro-electric power which is another alternative source of energy.

Formulate regulations to stop overfishing: Overfishing interrupts aquatic life and depletes the fish available in our water bodies. In some cases it poses a threat to the endangered aquatic species. Regulations to avoid over fishing should be put in place.

Control the number of cattle in a household to prevent overgrazing: Overgrazing destroys agricultural land making it prone to soil erosion.

Constructing terraces and gabions in sloping land: This will prevent soil erosion. We should always strive to protect the environment. We should therefore embrace these methods to protect the natural resources.

Reference

- Natural-resource-management, topic: britannica.com, Retrieved 13 March, 2020
- Thampapillai, Dodo J. (2002). Environmental economics: concepts, methods, and policies. Oxford University Press. ISBN 978-0-19-553577-8
- Terrific-ways-to-conserve-natural-resources: conserve-energy-future.com, Retrieved 15 July, 2020
- Gunderson, Lance H.; Holling, C.S., eds. (2002). Panarchy Synopsis: Understanding Transformations in Human and Natural Systems. Island Press. ISBN 978-1-55963-330-7
- Gis-and-natural-resource-management: gislounge.com, Retrieved 16 January, 2020
- "Evans, K., De Jong, W., and Cronkleton, P. (2008) "Future Scenarios as a Tool for Collaboration in Forest Communities". "S.A.P.I.EN.S." ""1"" (2)". Sapiens.revues.org. 1 October 2008. Retrieved 30 November2011
- Smil, Vaclav (2003). The Earth's Biosphere: Evolution, Dynamics, and Change. MIT Press. P. 25. ISBN 0-262-69298-8
- Willard, B. (2011). "3 Sustainability Models" citing The Power of Sustainable Thinking by Bob Doppelt, and The Necessary Revolution by Peter Senge et al. Retrieved on: 2011-05-03
- ASTM E60 "E2114-08 Standard Terminology for Sustainability", ASTM, 2008, pp. 615-618 ISBN 978-0-8031-5768-2
- Floyd, J.; Zubevich, K. (2010). "Linking foresight and sustainability: An integral approach". Futures. 42: 59. Doi:10.1016/j.futures.2009.08.001
- Sp-sustainable-ecosystem-management, reference-book, library: openness-project.eu, Retrieved 10 July, 2020

Water and Soil Conservation

4

- **Water Conservation**
- **Water Resource Management**
- **Rainwater Harvesting System**
- **Integrated Catchment Management**
- **Keyline Design**
- **Soil Conservation**

Water conservation refers to the activities and strategies towards the sustainable management of natural resources of water. The utilization of methods to prevent soil loss from erosion, acidification and salinization is known as soil conservation. The topics elaborated in this chapter will help in gaining a better perspective about the diverse aspects of water and soil conservation.

Water Conservation

Water conservation is the use and management of water for the good of all consumers. It is used in agriculture, industry, and the home. Human requirements for agricultural production, flood control, fish and wildlife management, navigation, industrial production, and many other uses have amended natural hydrologic processes.

The hydrosphere refers to that portion of Earth that is made of water, including all oceans, lakes, rivers, streams, glaciers, and underground water. Less than 3% of the water of Earth is freshwater, an amount that includes polar ice caps, glaciers, groundwater, surface water of rivers and freshwater lakes, and even atmospheric water. However, the amount of freshwater useable by people and other members of the biosphere is less than 0.7% of the total (this is water in rivers and lakes, and in the ground). This relatively small amount of available freshwater is recycled and purified by the action of processes within the hydrologic cycle, including evaporation, condensation, precipitation, and percolation through the ground. All life depends on the availability of freshwater.

Of all the freshwater used directly by humans, agricultural irrigation accounts for about 70% of the total. The remainder is used for industrial and domestic purposes. However, these proportions vary widely due to the climatic and economic conditions of the particular locality. Within this century, one-third of the countries situated in areas of water scarcity may encounter severe water shortages. By 2025, two thirds of the world's population is likely to live in areas of moderate or severe water shortage. The need for more effective conservation of the limited supplies of water that are available for use by people and required by natural ecosystems will intensify as water stress grows.

Freshwater Resources

Available freshwater resources are either ground-water or surface water (rivers and lakes). Water that flows on the surface of the land is surface runoff. The relationship among surface runoff, precipitation, evaporation, and percolation is summarized in the following equation:

$$\text{Surface runoff} = \text{precipitation} - (\text{evaporation} + \text{percolation})$$

When surface runoff resulting from rainfall or snowmelt is confined to a relatively narrow, well-defined channel, it is called a river or stream.

Groundwater is that water that has percolated downward through the soil and is present within porous spaces in soil and bedrock. It has been estimated that the global groundwater resource is equivalent to about 34 times the volume of all surface waters (i.e., rivers and lakes) of the world. This resource is present nearly everywhere and has the additional advantages of typically needing no storage or treatment. Utilization does require the construction of a well, sometimes presenting a problem in the most needy locations.

Water utilization efficiency is measured by the ratio of water withdrawal and its subsequent consumption. Water withdrawal is water pumped from rivers, reservoirs, or groundwater wells, and is then transported for use. Water consumption is water that is withdrawn and actually used for some specific purpose. It is then returned to the environment through evaporation, transpiration, discharge to a river or lake, or in some other way.

Water Consumption

Water consumption varies greatly among regions due to differences in economic development. The average municipal use in the United States is about 150 gal (568 l) per person per day, though the rate can be higher than 350 gal (1324 l) in some locations.

This includes home use for bathing, waste disposal, and gardening, as well as institutional and commercial usage. Per capita (per person) water usage in Asia is only 22 gal (85 l) per day, and just 12 gal (47 l) in Africa.

According to the World Health Organization (WHO) of the United Nations, people have a minimum water requirement of about 5 gal (20 l) per person per day. This is the minimum amount needed for physiological rehydration, cooking, washing, and other subsistence requirements. However, the WHO estimates that nearly two billion people consume contaminated water. This carries a significant risk of developing such water-borne diseases as cholera, dysentery, polio, or typhoid, which kill about 25 million people per year. Both conservation and sanitation are obvious necessities in meeting the huge demand for freshwater.

Because irrigation accounts for 70% of the water used by humans worldwide, achieving a better efficiency of agricultural use is a logical step in advancing water conservation. This can be accomplished by lining water delivery systems with concrete or other impervious materials to minimize loss by leaking during transport, and by using drip-irrigation systems to minimize losses by evaporation. Drip-irrigation systems have been successfully used on fruit trees, certain row-crops, and horticultural plants. Conservation can also be accomplished by improving the efficiency of utilization of water by crops, including the cultivation of plants that are less demanding of moisture.

Efficient Water Utilization Efforts

Subsurface irrigation is an emerging technology with high water-utilization efficiency. Subsurface irrigation uses a drip-irrigation tubing buried 6 to 8 in (15 to 20 cm) underground, with a spacing of 12 to 24 in (30 to 60 cm) between parallel lines. The tubing contains drip outlets that deliver water and nutrients within the root zone at a desired rate. In addition to water conservation, subsurface irrigation has other advantages that overhead sprinklers do not: minimal over watering, fewer disease and aeration problems, less runoff and erosion, fewer weeds, and better protection from vandalism. However, this system is relatively expensive to install. In California, subsurface irrigation has been used on fruit trees, field crops, and lawns, and has achieved water-use savings of about 50%. However, this methodology can, in arid environments, lead to the build-up of soil salinity levels, damaging plants and reducing crop yields. Balancing the water needs of the plant with maintenance of soil quality is an important component of water conservation measures. Technologically advanced irrigation systems now incorporate climate-based controls. These systems utilize meteorological information to determine the need for irrigation and modify the length and duration of irrigation to match the plant's requirements. Though these systems are currently used primarily on large-scale applications, development of economical models for the small-scale user is underway.

Xeriscaping, or the cultivation of plants requiring little water, is an especially suitable horticultural practice for conserving water in regions with a dry, hot climate. For example, over much of the southwestern United States, more than 50% of the domestic water consumption may be used to irrigate lawns and other horticultural plants that are intolerant of drought. Xeriscaping uses plants such as cacti, succulents, and shrubs of semi-desert habitat (such as trailing rosemary Rosemarinus officinale and rock rose Cistus cobariensis), which are well-adapted to a hot, dry climate and need little water.

Water conservation can also be advanced by improving other domestic uses of water. One simple conservation practice is to install ultra-low-flush (ULF) toilets and low-flow showerheads in homes and other buildings. A ULF toilet uses only 1.6 gal (6.1 l) per flush, compared to 5 to 7 gal by a standard toilet. Replacing a standard toilet with an ULF saves about 30 to 40 gal (114 to 151 l) of water per day, equivalent to 10,000 to 16,000 gal (37,850 to 60,560 l) per year. More recently, advanced toilets and urinals requiring no water have been developed and are beginning to be utilized on a limited basis.

Another way to conserve the freshwater supply is to desalinize seawater. Desalinization is the removal of salts and other impurities from seawater by either distillation or reverse osmosis (RO), and this method is being increasingly used to provide high-quality water for drinking, cooking, and other domestic uses. In 2004, the world production of desalinated water was at least 40 billion gallons per day (150 billion liters), most of which was produced in Saudi Arabia and other nations of the Gulf of Arabia, where energy costs are relatively low (the cost of desalinated water is highly sensitive to the cost of energy). The largest desalination plant in the world (Shoaiba Desalination Plant) is located in Saudi Arabia, and it uses reverse osmosis to produce half of its country's drinking water. Saudi Arabia is the largest producer of desalinated water in the world with desalination providing 70% of the country's drinking water. Desalinization is also practiced in California and Florida, where the cost is about three dollars per thousand gallons, which is four to five times the cost paid for domestic water by typical urban consumers in the United States, and more than 100 times the cost paid by farmers for water for irrigation. The process is also gaining popularity in Spain, Australia, and China.

Widespread recognition of the importance of reusing water has begun to change traditional water use methods. As the value of water increases, users are willing to employ methods that may increase the initial cost of a project, with the hope of regaining those costs through water savings in the future. One of the first of these reuse applications was the irrigation of golf courses and landscaping. In many areas, treated waste-water is diverted from its normal disposal path to be reused in irrigation. This has gained in popularity and is also utilized in small artificial ponds for decorative purposes. Graywater systems capture water that drains from sinks, tubs, laundry, and dishwashers for reuse in irrigation. Graywater systems do not incorporate toilet wastes because of the potential health threat. Dual plumbing is required for such a system and some treatment is required prior to reuse. Though home construction costs are obviously increased by including a graywater system, many have become dedicated believers in the benefits of water reuse, while others question the economic benefit of small-scale systems.

Economic Incentives for Water Conservation

As the availability of water becomes more restricted, the costs to both the provider and consumer are increased. In a situation unique to the water supply industry, providers are frequently placed in the position of trying to convince consumers to use less of the commodity that they supply. Most large water providers have departments dedicated to education of the public with regard to conservation. In general, these education efforts have been largely ineffective and conservation of freshwater resources has been best achieved through economic incentives. Water providers frequently provide rebates for those consumers that are willing to change from older technology to newer, such as low-flush toilets and modern washing machines, convert to water efficient landscaping, or otherwise demonstrate lower water usage. Greatest effect has been achieved through tiered pricing. In this pricing structure, users are charged higher rates for each successive unit, or block, of water used. The rate structure penalizes heavy users with greatly increased rates. This technique has been shown to be highly effective in reducing overall usage. In Tucson, Arizona, for example, an increasing tiered price structure resulted in decreased usage of 26% over a three-year period. Additionally, some communities have implemented the use of water conservation monitors and water waste hotlines to penalize those that continue to waste the resource. Many communities currently limit the type and size of landscaping, the time and nature of outdoor water use, and in extreme cases, have completely banned outdoor water use during crisis periods.

Throughout history, the availability of water has been a vital factor in the rise and fall of human cultures. This is largely because water is a limiting factor for the carrying capacity for human activities in any region. It is crucial that humans learn to live within the limits of available natural resources, including the supply of fresh water. Because the supply of usable water is finite, the consumption per person must be reduced in regions that are using this resource excessively.

Water Resource Management

Water resource management is the activity of planning, developing, distributing and managing the optimum use of water resources. It is a sub-set of water cycle management.

The field of water resources management will have to continue to adapt to the current and future issues facing the allocation of water. With the growing uncertainities of global climate change and the long term impacts of management actions,the decision-making will be even more difficult. It is likely that ongoing climate change will lead to situations that have not been encountered. As a result, alternative management strategies are sought for in order to avoid setbacks in the allocation of water resources.

Ideally, water resource management planning has regard to all the competing demands for water and seeks to allocate water on an equitable basis to satisfy all uses and demands. As with other resource management, this is rarely possible in practice.

One of the biggest concerns for our water-based resources in the future is the sustainability of the current and even future water resource allocation. As water becomes more scarce, the importance of how it is managed grows vastly. Finding a balance between what is needed by humans and what is needed in the environment is an important step in the sustainability of water resources. Attempts to create sustainable freshwater systems have been seen on a national level in countries such as Australia, and such commitment to the environment could set a model for the rest of the world.

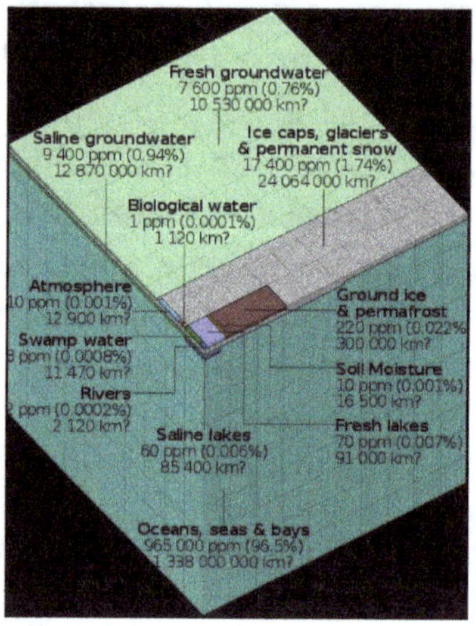

Visualisation of the distribution (by volume) of water on Earth. Each tiny cube (such as the one representing biological water) corresponds to approximately 1,000 cubic kilometres (240 cu mi) of water, with a mass of approximately 1 trillion tonnes (2000 times that of the Great Pyramid of Giza or 5 times that of Lake Kariba, arguably the heaviest man-made object). The entire block comprises 1 million tiny cubes.

Water is an essential resource for all life on the planet. Of the water resources on Earth only three percent of it is fresh and two-thirds of the freshwater is locked up in ice caps and glaciers. Of the remaining one percent, a fifth is in remote, inaccessible areas and much seasonal rainfall in monsoonal deluges and floods cannot easily be used. As time advances, water is becoming scarcer and having access to clean, safe, drinking water is limited among countries. At present only about 0.08 percent of all the world's fresh water is exploited by mankind in ever increasing demand for sanitation, drinking, manufacturing, leisure and agriculture. Due to the small percentage of water remaining, optimizing the fresh water we have left from natural resources has been a continuous difficulty in several locations worldwide.

Much effort in water resource management is directed at optimizing the use of water and in minimizing the environmental impact of water use on the natural environment. The observation of water as an integral part of the ecosystem is based on integrated water resource management, where the quantity and quality of the ecosystem help to determine the nature of the natural resources.

As a limited resource, water supply sometimes supposes a challenge. This fact is assumed by the project DESAFIO (the acronym for Democratisation of Water and Sanitation Governance by Means of Socio-Technical Innovations), which has been developed along 30 months and funded by the European Union's Seventh Framework Programme for research, technological development and demonstration. This project faced a difficult task for developing areas: eliminating structural social inequity in the access to indispensable water and public health services. The DESAFIO engineers worked on a water treatment system run with solar power and filters which provides safe water to a very poor community in the state of Minas Gerais.

Successful management of any resources requires accurate knowledge of the resource available, the uses to which it may be put, the competing demands for the resource, measures to and processes to evaluate the significance and worth of competing demands and mechanisms to translate policy decisions into actions on the ground.

For water as a resource, this is particularly difficult since sources of water can cross many national boundaries and the uses of water include many that are difficult to assign financial value to and may also be difficult to manage in conventional terms. Examples include rare species or ecosystems or the very long term value of ancient groundwater reserves.

Agriculture

Agriculture is the largest user of the world's freshwater resources, consuming 70 percent. As the world population rises it consumes more food, the industries and urban developments expand, and the emerging biofuel crops trade also demands a share of freshwater resources, water scarcity is becoming an important issue. An assessment of water resource management in agriculture was conducted in 2007 by the International Water Management Institute in Sri Lanka to see if the world had sufficient water to provide food for its growing population or not. It assessed the current availability of water for agriculture on a global scale and mapped out locations suffering from water scarcity. It found that a fifth of the world's people, more than 1.2 billion, live in areas of physical water scarcity, where there is not enough water to meet all their demands. A further 1.6 billion people live in areas experiencing economic water scarcity, where the lack of investment in water or insufficient human capacity make it impossible for authorities to satisfy the demand for water.

The report found that it would be possible to produce the food required in future, but that continuation of today's food production and environmental trends would lead to crises in many parts of the world. Regarding food production, the World Bank targets agricultural food production and water resource management as an increasingly global issue that is fostering an important and growing debate. The authors of the book Out of Water: From abundance to Scarcity and How to Solve the World's Water Problems, which laid down a six-point plan for solving the world's water problems. These are: 1) Improve data related to water; 2) Treasure the environment; 3) Reform water governance; 4) Revitalize agricultural water use; 5) Manage urban and industrial demand; and 6) Empower the poor and women in water resource management. To avoid a global water crisis, farmers will have to strive to increase productivity to meet growing demands for food, while industry and cities find ways to use water more efficiently.

Managing Water in Urban Settings

As the carrying capacity of the Earth increases greatly due to technological advances, urbanization in modern times occurs because of economic opportunity. This rapid urbanization happens worldwide but mostly in new rising economies and developing countries. Cities in Africa and Asia are growing fastest with 28 out of 39 megacities (a city or urban area with more than 10 million inhabitants) worldwide in these developing nations. The number of megacities will continue to rise reaching approximately 50 in 2025. With developing economies water scarcity is a very common and very prevalent issue. Global freshwater resources dwindle in the eastern hemisphere either than at the poles, and with the majority of urban development millions live with insufficient fresh water. This is caused by polluted freshwater resources, overexploited groundwater resources, insufficient harvesting capacities in the surrounding rural areas, poorly constructed and maintained water supply systems, high amount of informal water use and insufficient technical and water management capacities.

In the areas surrounding urban centres, agriculture must compete with industry and municipal users for safe water supplies, while traditional water sources are becoming polluted with urban runoff. As cities offer the best opportunities for selling produce, farmers often have no alternative to using polluted water to irrigate their crops. Depending on how developed a city's wastewater treatment is, there can be significant health hazards related to the use of this water. Wastewater from cities can contain a mixture of pollutants. There is usually wastewater from kitchens and toilets along with rainwater runoff. This means that the water usually contains excessive levels of nutrients and salts, as well as a wide range of pathogens. Heavy metals may also be present, along with traces of antibiotics and endocrine disruptors, such as oestrogens.

Developing world countries tend to have the lowest levels of wastewater treatment. Often, the water that farmers use for irrigating crops is contaminated with pathogens from sewage. The pathogens of most concern are bacteria, viruses and parasitic worms, which directly affect farmers' health and indirectly affect consumers if they eat the contaminated crops. Common illnesses include diarrhoea, which kills 1.1 million people annually and is the second most common cause of infant deaths. Many cholera outbreaks are also related to the reuse of poorly treated wastewater. Actions that reduce or remove contamination, therefore, have the potential to save a large number of lives and improve livelihoods. Scientists have been working to find ways to reduce contamination of food using a method called the 'multiple-barrier approach'.

This involves analysing the food production process from growing crops to selling them in markets and eating them, then considering where it might be possible to create a barrier against contamination. Barriers include: introducing safer irrigation practices; promoting on-farm wastewater treatment; taking actions that cause pathogens to die off; and effectively washing crops after harvest in markets and restaurants.

Urban Decision Support System

Urban Decision Support System (UDSS) – is a wireless device with a mobile app that uses sensors attached to water appliances in urban residences to collect data about water usage and is an example of data-driven urban water management. The system was developed with a European Commission investment of 2.46 Million Euros to improve the water consumption behaviour of households. Information about every mechanism – dishwashers, showers, washing machines, taps – is wirelessly recorded and sent to the UDSS App on the user's mobile device. The UDSS is then able to analyse and show homeowners which of their appliances are using the most water, and which behaviour or habits of the households are not encouraged in order to reduce the water usage, rather than simply giving a total usage figure for the whole property, which will allow people to manage their consumption more economically.

Rainwater Harvesting System

Rainwater harvesting system, also called rainwater collection system or rainwater catchment system is the technology that collects and stores rainwater for human use. Rainwater harvesting systems range from simple rain barrels to more elaborate structures with pumps, tanks, and purification systems. The nonpotable water can be used to irrigate landscaping, flush toilets, wash cars, or launder clothes, and it can even be purified for human consumption. With water scarcity a pressing problem for many densely populated regions, rainwater harvesting systems can supply households and businesses with water for use in dry seasons and lessen the demand on municipal systems.

Given that rainfall is sporadic and that only a small proportion of global precipitation is easily available for human use, rainwater harvesting can be an efficient means of capturing that precious resource. In cities, much of the rain that falls on buildings, roofs, roads, and other hard landscaping does not percolate into the soil and is instead directed into storm sewers for disposal. Impermeable surfaces cause urban flooding in many areas and generate contaminated unusable water that is directed away from potable water resources. During dry months, local groundwater can be depleted, and many localities struggle to consistently provide enough potable water to meet demand. Rainwater harvesting for nonpotable functions, such as gardening and washing clothes, significantly reduces both the demanded amount of the total fresh water and the strain on stormwater infrastructure. That saving in the demand and supply of potable fresh water is significant in large cities. Although many localities encourage and even subsidize rain barrels and other rainwater harvesting systems, some areas, particularly those in the southwestern United States, view rainwater harvesting as a water rights issue and place restrictions on such collections.

Design Recommendations

The simplest rainwater harvesting systems are nonpressurized systems, such as rain barrels, where the pipes run from rain gutters into a tank. Known as "dry systems," those structures do not hold any water in the pipes after it stops raining and do not create breeding grounds for mosquitoes and other insects. "Wet systems" are necessary when the pipes cannot be configured to run directly into the tanks. In places where the tanks are located some distance away from the collection surfaces or where there are a series of tanks to serve a number of buildings, pipes from the gutter go underground and then up through a riser into the tank. Such systems are often pressurized so that the long runs of pipes do not retain stagnant water.

Well-designed rainwater harvesting systems ensure that the pipes and all other openings are insect-proof, especially in wet systems. Additionally, wire mesh screen covers on all tank inlets can help prevent debris from entering the tank. Collection surfaces (mainly roofs) should be made of nontoxic materials, particularly avoiding lead-based paints and membranes, and tanks should be made of nontoxic and noncorrosive material. Care should be taken to ensure that the tank outlet taps or draw-off pipes are at least 10 cm (4 inches) above the tank floor to avoid drawing out any sludge that may have collected in the water supply. Although some systems have a sump pump and washout pipe to remove sludge, rqualityegular cleaning of the inside surfaces of the tank is recommended for all systems.

Additionally, catchments should be kept clean of accumulations of dirt, moss, lichens, and other debris. Trees branches that overhang those catchment surfaces should be cut back. Regular cleaning of gutters, tank inlets, and screens and annual tank inspection are necessary for proper functioning. Ideally, the water should be tested periodically to monitor its.

Quality

Rainwater mixes with both soluble and insoluble materials from the surfaces on which it lands and collects dust and pollutants as it flows down through the atmosphere. Contaminants may be plants, fungi, and other organic materials, as well as inorganic substances such as dissolved minerals, metals, chemicals, or water-soluble paints. Although collected rainwater does not need a high degree of purity for garden or agricultural applications, rainwater collected from unclean surface runoffs is not suitable for drinking or cooking. Separation of the first flush of rainwater from the roof, gutters, and other collection surfaces can improve water quality in the rainwater storage tank.

If collected rainwater is intended for household uses, it must first be purified. Flocculation, settlement, and biofilm skimming can be used to remove bacteria, organic material, and chemicals that form films on surfaces or settle to the bottom of the tanks as sludge. A liquid alum solution can also be added to the incoming raw water to bind fine suspended particles to form larger particles that can be removed by settling and filtration. That removes objectionable colour, turbidity (cloudiness), and aluminum from the drinking water. Prefiltered water may then undergo solar water disinfection or be treated with chlorine or other chemicals to kill infectious agents if the supply is intended for potable uses. Other chemicals used for water purification are potassium permanganate, calcium hydroxide, and fluoride.

Integrated Catchment Management

Effective management of water requires us to look at the pressures on our water resources at an appropriate scale - large enough that we can take account of all the relevant information, but small enough to ensure that people who live in the area can easily relate to their catchment. Experience around the world has shown that an integrated approach to managing individual catchments of an appropriate scale is necessary to protect and improve water resources. A catchment is simply defined as an area contributing water to a river and its tributaries, with all the water ultimately running off to a single outlet.

3D depiction of a catchment showing the water cycle.

- Gathering the best available information to understand the catchment - where the water comes from, how it flows through the landscape both overground and underground, and what activities in the catchment may be causing pollution.
- Looking at all the uses of water - drinking, agricultural, industrial and recreational, and also the vital ecosystems that depend on water to survive.

- Engaging local communities and involving them in decision making and management of their catchment.
- Adopting appropriate measures to ensure that activities that represent a significant threat to water resources are effectively managed.
- Applying the scientific and local knowledge of how the catchment operates to protect and improve water, providing a healthy, resilient, productive and valued resource that supports vibrant communities.

An Integrated Catchment Management approach is increasingly seen internationally as essential to successful water management.

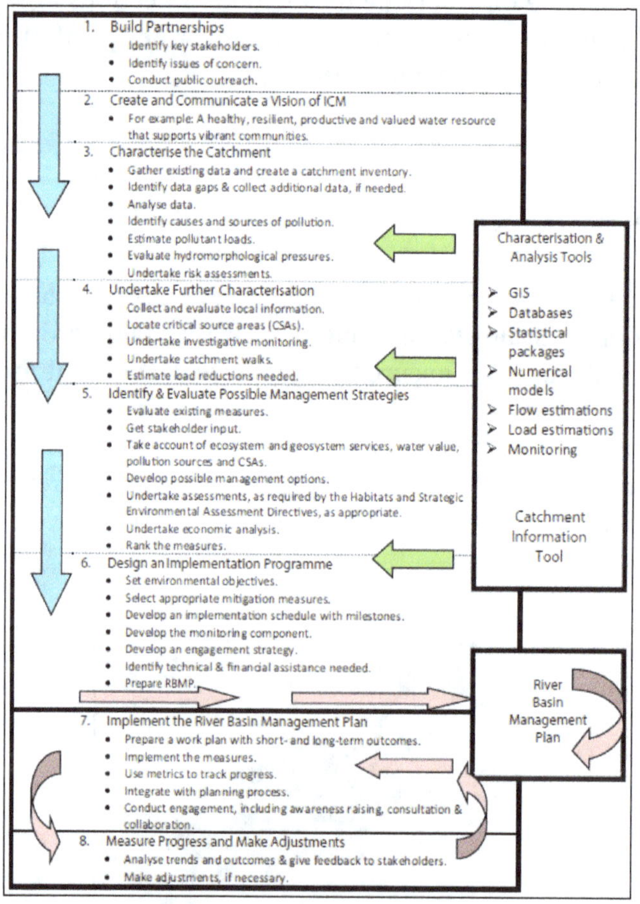

Steps in the ICM Process.

Keyline Design

Keyline design is a landscaping technique of maximizing the beneficial use of the water resources of a tract of land. The "keyline" denominates a specific topographic feature related to the natural flow of water on the tract. Keyline design is a system of principles and techniques of developing rural and urban landscapes to optimize use of their water resources.

Application

P. A. Yeomans published the first book on Keyline design in 1954. Yeomans described a system of amplified contour ripping to control rainfall runoff and enable fast flood irrigation of undulating land without the need for terracing it.

Keyline designs include irrigation dams equipped with through-the-wall lockpipe systems, gravity feed irrigation, stock water, and yard water. Graded earthen channels may be interlinked to broaden the catchment areas of high dams, conserve the height of water, and transfer rainfall runoff into the most efficient high dam sites. Roads follow both ridge lines and water channels to provide easier movement across the land.

Keyline Scale of Permanence

The foundation of Yeomans' Keyline design system is the Keyline Scale of Permanence (KSOP), which was the outcome of 15 years of adaptive experimentation. The Scale identifies the environmental elements of typical farms and orders them according to their degree of permanence as follows:

- Climate,
- Landshape (topography),
- Water supply,
- Roads and means of access,
- Trees,
- Structures (edifices),
- Subdivisional fences,
- Soil.

Keyline design considers these elements in planning the placement of water storage features, roads, trees, edifices, and fences. On undulating land, keyline design involves identifying ridges, valleys, and natural water courses and designing with them in mind in order to optimize water storage sites. Constructing interconnecting channels may be part of such optimization.

The identified natural water lines delineate the possible locations for the various less permanent elements, e. g. roads, fences, trees, and edifices, which if so located would help optimize the natural potential of the land in question.

Rancho San Ricardo, Oaxaca, México.

Keypoint

In a smooth, grassy valley, a location denominated the keypoint is identified at which the lower and leveller portion of the primary valley floor suddenly steepens higher. The keyline of this primary valley is determined by pegging a contour line that conforms to the natural shape of the valley through the keypoint, such that all points on the keyline are at the same elevation as the keypoint. Contour plowing both above and below the keyline and parallel to it ipso facto is "off-contour", but the developing pattern tends to drift rainwater runoff away from the center of the valley and, incidentally, prevent erosion of its soil.

Cultivation conforming to Keyline design for ridges is done parallel to any suitable contour, but only on the high side of the contour's guide line. This ipso facto develops a pattern of off-contour cultivation in which all the rip marks made in the soil slope down towards the center of the ridge. This pattern of cultivation allows more time for water to infiltrate. Cultivation following Keyline pattern also enables controlled flood irrigation of undulating land, which increases the rate of development of deep, fertile soil.

In many nations, including Australia, it is important to optimize infiltration of rainfall, and Keyline cultivation accomplishes this while delaying the concentration of runoff that could damage the land. Yeomans' technique differs from traditional contour plowing in several important respects. Random contour plowing also becomes off contour but usually with the opposite effect on runoff, namely causing it to quickly run off ridges and concentrate in valleys. The limitations of the traditional system of soil conservation, with its "safe disposal" approach to farm water, was an important motivation to develop Keyline design.

Applications

David Holmgren, one of the founders of permaculture, used Yeoman's Keyline design extensively in the formulation of his principles of permaculture and the design of sustainable human settlements and organic farms.

Darren J. Doherty has extensive global experience in Keyline design, development, management, and education. He uses Keyline as the basis for his Regrarians framework, which he considers a revision and synthesis of Keyline design, permaculture, holistic management, and several other innovative, human ecological frameworks into a coherent process-based system of design and management of regenerative economies.

A topographical example of Keyline design is available at (37.159154°S 144.252248°E).

Keyline design also includes principles of rapidly enhancing soil fertility. They are explored in *Priority One* by P. A. Yeomans' son, Allan. Yeomans and his sons were also instrumental in the design and production of special plows and other equipment for Keyline cultivation.

Soil Conservation

Soil Conservation is the name given to a handful of techniques aimed at preserving the soil. Soil loss and loss of soil fertility can be traced back to a number of causes including over-use, erosion, salinization and chemical contamination. Unsustainable subsistence farming and the slash and burn clearing methods used in some less developed regions, can often cause deforestation, loss of soil nutrients, erosion on a massive scale and sometimes even complete desertification.

Soil erosion removes the top soil that is necessary for organic matter, nutrients, microorganisms that are requires for plants to grow and shine. Soil conservation is one such step that protects the soil from being washed away. The soil then ends up in aquatic resources bringing in pesticides and fertilizers used on agricultural land. Healthy soil is important for plants to grow and flourish. Taking necessary steps to conserve the soil is part of environmentally friendly lifestyle. There are several ways to conserve soil that can be done through agricultural practices or measures you take at home.

Methods and Techniques of Soil Conservation

Many different techniques have been invented throughout the years with the aim of preserving the nutrient level of the soil and preventing erosion.

- Contour Plowing: Originating in ancient Phoenicia, Contour Plowing involves plowing grooves into the desired farmland, then planting the crop furrows in the grooves and following the contours. It a very effective way for farmland on slopes to prevent run off improve crop yields.
- Terrace Farming: Terracing is a method of carving multiple, flat leveled areas into hills. Steps are formed by the terraces which are surrounded by a mud wall to prevent run off and hold the soil nutrients in the beds. More commonly found in lesser developed nations due to the difficulty of using mechanized farming equipment in the terraces. Very popular in Asia for planting rice.
- Keyline Design: A more enhanced version of Contour Plowing, maximum water retention is achieved by taking into account all the watershed properties when making the contour lines. The Keyline refers to topographic feature linked to water flow. This allows the water run off to run directly into an existing water channel, and prevent soil erosion caused by the water.
- Perimeter Runoff Control: This is the practice of planting trees, shrubs and ground cover around the perimeter of your farmland which impedes surface flows and keeps nutrients in the farmed soil. Using the grass way is a specialized way of handling perimeter runoff that uses surface friction to channel and dissipate runoff.
- Windbreaks: Rows of tall trees are used in dense patterns around the farmland and prevents wind erosion. Evergreen trees can provide year round protection but deciduous trees can be adequate as long as foliage is apparent during the seasons when the soil is bare.
- Cover Crops/Crop Rotation: Cover crops such as turnips and radishes are rotated with cash crops in order to blanket the soil all year- round and produces green manure the replenishes nitrogen and other critical nutrients. Using cover crops can also suppress weeds.
- Soil Conservation Farming: A mixture of farming methods intending the mimic the biology of virgin land. These practices can be used to prevent erosion and even restore damaged soil and encourage plant growth. Eliminating the use of nitrogen fertilizer and fungicides can increase yields and protect crops from drought and flooding.
- Agrostological Measures: Planting grass in heavily eroded areas is called an agrostological measure. Ley farming practices cultivating grass in rotation with regular crops to increase the nutrient level in the soils. When the grass is

harvested it can be used as fodder for cattle. For heavily eroded soil it is recommended to grown grass for many years to let the soils naturally repair themselves.

- No till farming: This is the method of growing crops year round without changing the topography of the soil by tilling or contouring. This technique increases the amount of water that penetrates the soil and can increase organic matter of the soil which leads to larger yields.
- Green Manures: Green manures are a few different crops that can be grown, not for produce or food usage, but grown in order to fertilize the farm land on which it grows. This method can improve the soil structure and suppresses the growth of weeds.
- Salinity Management: When water evaporates from the soil, it leaves behind its salt. This can lead to damage of the soil and nutrient loss. Using humic acids can prevent this or growing crops like saltbush can rejuvenate the soils and replace lost nutrients. High levels of salt in the soil can often be caused by changes made to the water table by damming and other causes.
- Stream Bank Protection: During floods, stream banks can often cave in. Preventing this by constructing walls along the banks or plant useful tree species will prevent this in the future and prevent soil loss down the stream.
- Earthworms: Earthworms provide great benefits of farm land due to the way they burrow under the ground and provide more are for water to rest after it has infiltrated the soil. When these worms excrete egesta, this sits in the soils and gives the crops many nutrients which are absorbed via the roots of the plants. Earthworm casts contain a vast amount more nutrients than any natural soil in the world, and for that reason should be invited into the soils of farmland to help prevent erosion and will lead to larger crop yields.
- Mineralization: To help crops reach higher yields or full potential, sometimes crushed rock or chemical supplements are added to the farmland, this helps combat mineral depletion. Normally used after flooding, it brings substantial amounts of sediment which can damage the nutrient level of the soil.
- Korean Natural Farming: This method takes advantage of natural and indigenous microorganisms to produce fertile soils that yield high output and gets rid of the need to use herbicides or pesticides. An improvement in soil health and output is what keeps this method used in the respective areas.
- Reduction of Impervious Surfaces: Driveways patios and paved pathways allow precipitation to flow freely off them. As the water flows it picks up momentum and in turn erodes any soil in which it flows over after leaving the impervious surfaces, reducing the amount of these around your farmland will prevent erosion.

- Dry Farming: In areas with a very low amount of rainfall, crops which require very little water should be grown, this will lead to the preservation of the natural levels of moisture and nutrients in the soil.
- Rain Gardens: A rain garden is a shallow depression in the land which holds and collects running water from impervious surfaces and prevents erosion while saving the nutrients that inevitably get washed away. This also gives you a good bed to grow wetland plants.
- Re-establish Forest Cover: A dense amount of trees in a forest leads to a vast network of deep roots that offer a long term solution to soil erosion, another benefit is the windbreak that these trees can provide.
- Maintaining PH levels of soil: Contamination of soils due to acid rains and other pollutants can lead to loss of soil fertility. Use a Ph indicator monthly to check the levels of acids in the soil and treat the soils with eco-friendly chemicals to prevent a loss of crops and low yields.
- Indigenous Crops: The growth of indigenous crops is a good way to conserve soil, as the plants have a natural need for the nutrients in the soil in your area, they help to prevent soil erosion. If you grow non indigenous crops a recommended tip is to plant indigenous crops around them, in order to prevent soil erosion.
- Prevent Overgrazing: Try not to let overgrazing happen by moving herds around often. If overgrazing occurs, plant hardier and more nutritious species of forage in order to rebuild the soil. You can also harvest these crops and feed them to the grazers during the winter season.
- The Sharing of Knowledge: More developed countries can and should share their farming knowledge gained throughout the years with the lesser developed nations of the world. This will lead to a better quality of soil worldwide and can help to prevent famine and solves the food crisis in some areas of the world.

Reference

- Water-conservation, scandinavian-literature-biographies, biographies, literature-and-arts: encyclopedia.com, Retrieved 12 July, 2020
- The World Bank, 2006 "Reengaging in Agricultural Water Management: Challenges and Options". Pp. 4–5. Retrieved 2011-10-30
- Methods of soil-conservation: conserve-energy-future.com, Retrieved 1 August, 2020
- "Integrated Support System for Efficient Water Usage and Resources Management". Issewatus.eu. Retrieved 2017-01-10.
- Rainwater-harvesting-system, technology: britannica.com, Retrieved 11 January, 2020

Environmental Pollution Management

5

- **Pollution Control**
- **Air Pollution Control**
- **Sewage Treatment**
- **Industrial Wastewater Treatment**
- **Noise Pollution Control**

Environmental pollution management refers to the control of emissions and effluents into soil, air and water. It includes the areas of air pollution control, sewage treatment, industrial wastewater treatment and noise pollution control. This chapter closely examines these key concepts of environmental pollution management to provide an extensive understanding of the subject.

Pollution Control

Pollution control, in environmental engineering is any of a variety of means employed to limit damage done to the environment by the discharge of harmful substances and energies. Specific means of pollution control might include refuse disposal systems such as sanitary landfills, emission control systems for automobiles, sedimentation tanks in sewerage systems, the electrostatic precipitation of impurities from industrial gas, or the practice of recycling.

Next to the conservation of species from the loss of biological diversity, the control of pollution is the conservation problem of greatest magnitude; it might even be argued that pollution control is more urgent and important. Ultimately, the control of pollution involves a number of social decisions: 1) not to allow the escape into the environment of substances or forms of energy that are harmful to life, 2) to contain and recycle those substances that could be harmful if released into the environment in excessive quantities, and 3) not to release into the environment substances that persist and are toxic to living things. The knowledge and technology needed to put these decisions to work are now available. Pollution control does not mean an abandonment of existing productive human activities but their reordering so as to guarantee that their side effects do not outweigh their advantages.

However, for economic reasons, none of these measures is applied universally, and political and social pressures have not yet forced their application. Developing countries have expressed fear that excessive concern over pollution could impede their economic development—and indeed some of these countries have become sanctuaries for industries that find it less expensive to operate there than in areas with more rigorous standards. It is apparent that pollution control, regardless of the advanced state of its technology, will become a reality only when people demand it and only when nations are willing to agree on appropriate international standards.

Air Pollution Control

Air pollution control are the techniques employed to reduce or eliminate the emission into the atmosphere of substances that can harm the environment or human health. The control of air pollution is one of the principal areas of pollution control, along with wastewater treatment, solid-waste management, and hazardous-waste management.

Air is considered to be polluted when it contains certain substances in concentrations high enough and for durations long enough to cause harm or undesirable effects. These include adverse effects on human health, property, and atmospheric visibility. The atmosphere is susceptible to pollution from natural sources as well as from human activities. Some natural phenomena, such as volcanic eruptions and forest fires, may have not only local and regional effects but also long-lasting global ones. Nevertheless, only pollution caused by human activities, such as industry and transportation, is subject to mitigation and control.

Most air contaminants originate from combustion processes. During the Middle Ages the burning of coal for fuel caused recurrent air pollution problems in London and other large European cities. Beginning in the 19th century, in the wake of the Industrial Revolution, increasing use of fossil fuels intensified the severity and frequency of air pollution episodes. The advent of mobile sources of air pollution—i.e., gasoline-powered highway vehicles—had a tremendous impact on air quality problems in cities. It was not until the middle of the 20th century, however, that meaningful and lasting attempts were made to regulate or limit emissions of air pollutants from stationary and mobile sources and to control air quality on both regional and local scales.

The primary focus of air pollution regulation in industrialized countries has been on protecting ambient, or outdoor, air quality. This involves the control of a small number of specific "criteria" pollutants known to contribute to urban smog and chronic public health problems. The criteria pollutants include fine particulates, carbon monoxide, sulfur dioxide, nitrogen dioxide, ozone, and lead. Since the end of the 20th century, there also has been a recognition of the hazardous effects of trace amounts of many other air pollutants called "air toxics." Most air toxics are organic chemicals, comprising molecules that contain carbon, hydrogen, and other atoms. Specific emission regulations have been implemented against those pollutants. In addition, the long-term and far-reaching effects of the "greenhouse gases" on atmospheric chemistry and climate have been observed, and cooperative international efforts have been undertaken to control those pollutants. The greenhouse gases include carbon dioxide, chlorofluorocarbons (CFCs), methane, nitrous oxide, and ozone. In 2009 the U. S. Environmental Protection Agency ruled that greenhouse gases posed a threat to human health and could be subject to regulation as air pollutants.

The best way to protect air quality is to reduce the emission of pollutants by changing to cleaner fuels and processes. Pollutants not eliminated in this way must be collected or trapped by appropriate air-cleaning devices as they are generated and before they can escape into the atmosphere.

Control of Particulates

Airborne particles can be removed from a polluted airstream by a variety of physical processes. Common types of equipment for collecting fine particulates include cyclones, scrubbers, electrostatic precipitators, and baghouse filters. Once collected, particulates adhere to each other, forming agglomerates that can readily be removed from the equipment and disposed of, usually in a landfill.

Because each air pollution control project is unique, it is usually not possible to decide in advance what the best type of particle-collection device (or combination of devices) will be; control systems must be designed on a case-by-case basis. Important particulate characteristics that influence the selection of collection devices include corrosivity, reactivity, shape, density, and especially size and size distribution (the range of different particle sizes in the airstream). Other design factors include airstream characteristics (e.g., pressure, temperature, and viscosity), flow rate, removal efficiency requirements, and allowable resistance to airflow. In general, cyclone collectors are often used to control industrial dust emissions and as pre-cleaners for other kinds of collection devices. Wet scrubbers are usually applied in the control of flammable or explosive dusts or mists from such sources as industrial and chemical processing facilities and hazardous-waste incinerators; they can handle hot airstreams and sticky particles. Electrostatic precipitators and fabric-filter baghouses are often used at power plants.

Cyclones

A cyclone removes particulates by causing the dirty airstream to flow in a spiral path inside a cylindrical chamber. Dirty air enters the chamber from a tangential direction at the outer wall of the device, forming a vortex as it swirls within the chamber. The larger particulates, because of their greater inertia, move outward and are forced against the chamber wall. Slowed by friction with the wall surface, they then slide down the wall into a conical dust hopper at the bottom of the cyclone. The cleaned air swirls upward in a narrower spiral through an inner cylinder and emerges from an outlet at the top. Accumulated particulate dust is periodically removed from the hopper for disposal.

Cyclones are best at removing relatively coarse particulates. They can routinely achieve efficiencies of 90 percent for particles larger than about 20 micrometres (μm; 20 millionths of a metre). By themselves, however, cyclones are not sufficient to meet stringent air quality standards. They are typically used as pre-cleaners and are followed by more efficient air-cleaning equipment such as electrostatic precipitators and baghouses.

Scrubbers

Devices called wet scrubbers trap suspended particles by direct contact with a spray of water or other liquid. In effect, a scrubber washes the particulates out of the dirty airstream as they collide with and are entrained by the countless tiny droplets in the spray.

Several configurations of wet scrubbers are in use. In a spray-tower scrubber, an upward-flowing airstream is washed by water sprayed downward from a series of nozzles. The water is recirculated after it is sufficiently cleaned to prevent clogging of the nozzles. Spray-tower scrubbers can remove 90 percent of particulates larger than about 8 μm.

In orifice scrubbers and wet-impingement scrubbers, the air-and-droplet mixture collides with a solid surface. Collision with a surface atomizes the droplets, reducing droplet size and thereby increasing total surface contact area. These devices have the advantage of lower water-recirculation rates, and they offer removal efficiencies of about 90 percent for particles larger than 2 μm.

Venturi scrubbers are the most efficient of the wet collectors, achieving efficiencies of more than 98 percent for particles larger than 0.5 μm in diameter. Scrubber efficiency depends on the relative velocity between the droplets and the particulates. Venturi scrubbers achieve high relative velocities by injecting water into the throat of a venturi channel—a constriction in the flow path—through which particulate-laden air is passing at high speed.

Electrostatic Precipitators

Electrostatic precipitation is a commonly used method for removing fine particulates from airstreams. In an electrostatic precipitator, particles suspended in the airstream are given an electric charge as they enter the unit and are then removed by the influence of an electric field. The precipitation unit comprises baffles for distributing airflow, discharge and collection electrodes, a dust clean-out system, and collection hoppers. A high voltage of direct current (DC), as much as 100,000 volts, is applied to the discharge electrodes to charge the particles, which then are attracted to oppositely charged collection electrodes, on which they become trapped.

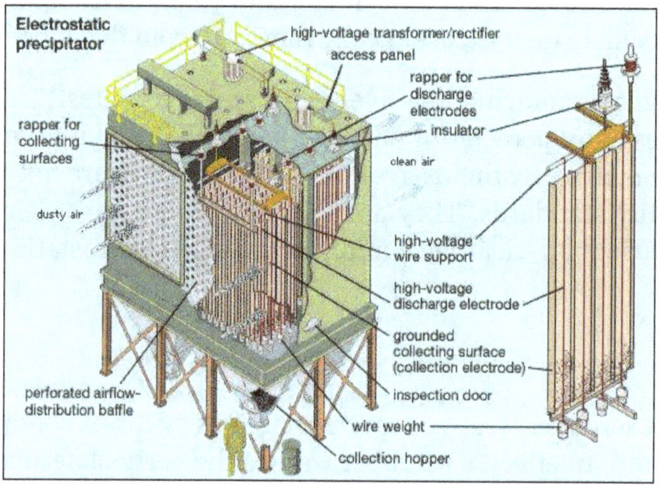

Electrostatic precipitator, a common particle-collection device at fossil-fuel power-generating stations.

In a typical unit the collection electrodes comprise a group of large rectangular metal plates suspended vertically and parallel to each other inside a boxlike structure. There are often hundreds of plates having a combined surface area of tens of thousands of square metres. Rows of discharge electrode wires hang between the collection plates. The wires are given a negative electric charge, whereas the plates are grounded and thus become positively charged.

Particles that stick to the collection plates are removed periodically when the plates are shaken, or "rapped." Rapping is a mechanical technique for separating the trapped particles from the plates, which typically become covered with a 6-mm (0.2-inch) layer of dust. Rappers are either of the impulse (single-blow) or vibrating type. The dislodged particles are collected in a hopper at the bottom of the unit and removed for disposal. An electrostatic precipitator can remove particulates as small as 1 μm with an efficiency exceeding 99 percent. The effectiveness of electrostatic precipitators in removing fly ash from the combustion gases of fossil-fuel furnaces accounts for their high frequency of use at power stations.

Baghouse Filters

One of the most efficient devices for removing suspended particulates is an assembly of fabric-filter bags, commonly called a baghouse. A typical baghouse comprises an array of long, narrow bags—each about 25 cm (10 inches) in diameter—that are suspended upside down in a large enclosure. Dust-laden air is blown upward through the bottom of the enclosure by fans. Particulates are trapped inside the filter bags, while the clean air passes through the fabric and exits at the top of the baghouse.

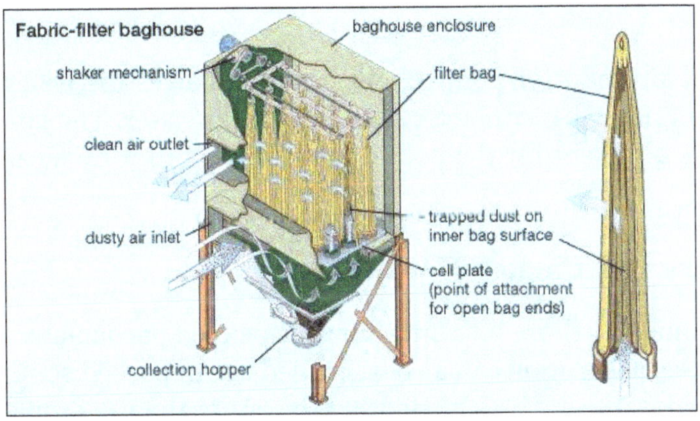

Baghouse employing an array of fabric bags for filtering the airstream.

A fabric-filter dust collector can remove very nearly 100 percent of particles as small as 1 μm and a significant fraction of particles as small as 0.01 μm. Fabric filters, however, offer relatively high resistance to airflow, which leads to substantial energy usage for the fan system. In addition, in order to prolong the useful life of the filter fabric, the air to be cleaned must be cooled (usually below 300 °C [570 °F]) before it is passed through the unit; cooling coils needed for this purpose add to the energy usage. (Certain filter fabrics—e.g., those made of ceramic or mineral materials—can operate at higher temperatures).

Several compartments of filter bags are often used at a single baghouse installation. This arrangement allows individual compartments to be cleaned while others remain in service. The bags are cleaned by removing the excess layer of surface dust. This is done in several different ways: by mechanically shaking them; by temporarily reversing the flow of air and causing them to collapse; or by sending a short burst of air down through the bag, causing it to briefly expand. After the dust is removed from the filters, it falls into a hopper below and can be collected for disposal or further use. Care must be taken not to remove too much of the built-up surface dust, or "dust cake," when cleaning the filters. In most filter types the filter itself is only a substrate that allows for the formation of a layer of dust cake, which then captures the majority of the particulates. Filters with an applied membrane coating such as polytetrafluoroethylene (Teflon) do not require the use of dust cake to operate at their highest efficiency.

Control of Gases

Gaseous criteria pollutants, as well as volatile organic compounds (VOCs) and other gaseous air toxics, are controlled by means of three basic techniques: absorption, adsorption, and incineration (or combustion). These techniques can be employed singly or in combination. They are effective against the major greenhouse gases as well. In addition, a fourth technique, known as carbon sequestration, is in development as a means of controlling carbon dioxide levels.

Absorption

In the context of air pollution control, absorption involves the transfer of a gaseous pollutant from the air into a contacting liquid, such as water. The liquid must be able either to serve as a solvent for the pollutant or to capture it by means of a chemical reaction.

Wet Scrubbers and Packed Scrubbers

Wet scrubbers similar to those for controlling suspended particulates may be used for gas absorption. Gas absorption can also be carried out in packed scrubbers, or towers, in which the liquid is present on a wetted surface rather than as droplets suspended in the air. A common type of packed scrubber is the countercurrent tower. After entering the bottom of the tower, the polluted airstream flows upward through a wetted column of light, chemically inactive packing material. The liquid absorbent flows downward and is uniformly spread throughout the column packing, thereby increasing the total area of contact between gas and liquid. Thermoplastic materials are most widely used as packing for countercurrent scrubber towers. These devices usually have gas-removal efficiencies of 90–95 percent.

Cocurrent and cross-flow packed scrubber designs are also used for gas absorption. In the cocurrent design, both gas and liquid flow in the same direction—vertically downward through the scrubber. Although not as efficient as countercurrent designs, cocurrent devices can work at higher liquid flow rates. The increased flow prevents plugging of the packing when the airstream contains high levels of particulates. Cocurrent designs afford lowered resistance to airflow and allow the cross-sectional area of the tower to be reduced. The cross-flow design, in which gas flows horizontally through the packing and liquid flows vertically downward, can operate with lower airflow resistance when high particulate levels are present.

In general, scrubbers are used at fertilizer production facilities (to remove ammonia from the airstream), at glass production plants (to remove hydrogen fluoride), at chemical plants (to remove water-soluble solvents such as acetone and methyl alcohol), and at rendering plants (to control odours).

Flue Gas Desulfurization

Sulfur dioxide in flue gas from fossil-fuel power plants can be controlled by means of an absorption process called flue gas desulfurization (FGD). FGD systems may involve wet scrubbing or dry scrubbing. In wet FGD systems, flue gases are brought in contact with an absorbent, which can be either a liquid or a slurry of solid material. The sulfur dioxide dissolves in or reacts with the absorbent and becomes trapped in it. In dry FGD systems, the absorbent is dry pulverized lime or limestone; once absorption occurs, the solid particles are removed by means of baghouse filters. Dry FGD systems, compared with wet systems, offer cost and energy savings and easier operation, but they require higher chemical consumption and are limited to flue gases derived from the combustion of low-sulfur coal.

FGD systems are also classified as either regenerable or nonregenerable (throwaway), depending on whether the sulfur that is removed from the flue gas is recovered or discarded. In the United States most systems in operation are nonregenerable because of their lower capital and operating costs. By contrast, in Japan regenerable systems are used extensively, and in Germany they are required by law. Nonregenerable FGD systems produce a sulfur-containing sludge residue that requires appropriate disposal. Regenerable FGD systems require additional steps to convert the sulfur dioxide into useful by-products like sulfuric acid.

Several FGD methods exist, differing mainly in the chemicals used in the process. FGD processes that employ either lime or limestone slurries as the reactants are widely applied. In the limestone scrubbing process, sulfur dioxide reacts with limestone (calcium carbonate) particles in the slurry, forming calcium sulfite and carbon dioxide. In the lime scrubbing process, sulfur dioxide reacts with slaked lime (calcium hydroxide), forming calcium sulfite and water. Depending on sulfur dioxide concentrations and oxidation conditions, the calcium sulfite can continue to react with water, forming calcium sulfate (gypsum). Neither calcium sulfite nor calcium sulfate is very soluble in water, and both can be precipitated out as a slurry by gravity settling. The thick slurry, called FGD sludge, creates a significant disposal problem. Flue gas desulfurization helps to reduce ambient sulfur dioxide levels and mitigate the problem of acid rain. Nevertheless, in addition to its expense (which is passed on directly to the consumer as higher rates for electricity), millions of tons of FGD sludge are generated each year.

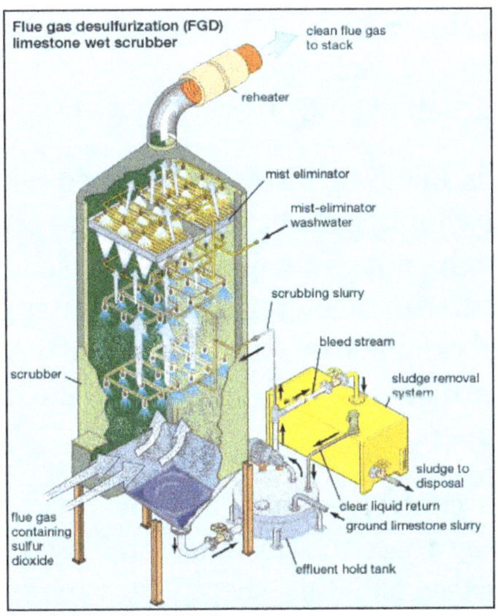

Wet scrubber using a limestone slurry to remove sulfur dioxide from flue gas.

Adsorption

Gas adsorption, as contrasted with absorption, is a surface phenomenon. The gas molecules are sorbed—attracted to and held—on the surface of a solid. Gas adsorption methods are used for odour control at various types of chemical-manufacturing and food-processing facilities, in the recovery of a number of volatile solvents (e.g., benzene), and in the control of VOCs at industrial facilities.

Activated carbon (heated charcoal) is one of the most common adsorbent materials. It is very porous and has an extremely high ratio of surface area to volume. Activated carbon is particularly useful as an adsorbent for cleaning airstreams that contain VOCs and for solvent recovery and odour control. A properly designed carbon adsorption unit can remove gas with an efficiency exceeding 95 percent.

Adsorption systems are configured either as stationary bed units or as moving bed units. In stationary bed adsorbers, the polluted airstream enters from the top, passes through a layer, or bed, of activated carbon, and exits at the bottom. In moving bed adsorbers, the activated carbon moves slowly down through channels by gravity as the air to be cleaned passes through in a cross-flow current.

Incineration

The process called incineration or combustion—chemically, rapid oxidation—can be used to convert VOCs and other gaseous hydrocarbon pollutants to carbon dioxide and water. Incineration of VOCs and hydrocarbon fumes usually is accomplished in a special incinerator called an afterburner. To achieve complete combustion, the afterburner must provide the proper amount of turbulence and burning time, and it must maintain a sufficiently high temperature. Sufficient turbulence, or mixing, is a key factor in combustion because it reduces the required burning time and temperature. A process called direct flame incineration can be used when the waste gas is itself a combustible mixture and does not need the addition of air or fuel.

An afterburner typically is made of a steel shell lined with refractory material such as firebrick. The refractory lining protects the shell and serves as a thermal insulator. Given enough time and high enough temperatures, gaseous organic pollutants can be almost completely oxidized, with incineration efficiency approaching 100 percent. Certain substances, such as platinum, can act in a manner that assists the combustion reaction. These substances, called catalysts, allow complete oxidation of the combustible gases at relatively low temperatures.

Afterburners are used to control odours, destroy toxic compounds, or reduce the amount of photochemically reactive substances released into the air. They are employed at a variety of industrial facilities where VOC vapours are emitted from combustion processes or solvent evaporation (e.g., petroleum refineries, paint-drying facilities, and paper mills).

Carbon Sequestration

The best way to reduce the levels of carbon dioxide in the air is to use energy more efficiently and to reduce the combustion of fossil fuels by using alternative energy sources (e.g., nuclear, wind, tidal, and solar power). In addition, carbon sequestration can be used to serve the purpose. Carbon sequestration involves the long-term storage of carbon dioxide underground, as well as on the surface of Earth in forests and oceans. Carbon sequestration in forests and oceans relies on natural processes such as forest growth. However, the clearing of forests for agricultural and other purposes (and also the pollution of oceans) diminishes natural carbon sequestration. Storing carbon dioxide underground—a technology under development that is also called geosequestration or carbon capture and storage—would involve pumping the gas directly into underground geologic "reservoir" layers. This would require the separation of carbon dioxide from power plant flue gases (or some other source)—a costly process.

Sewage Treatment

Sewage treatment is the process of removing contaminants from municipal wastewater, containing mainly household sewage plus some industrial wastewater. Physical, chemical, and biological processes are used to remove contaminants and produce treated wastewater (or treated effluent) that is safe enough for release into the environment. A by-product of sewage treatment is a semi-solid waste or slurry, called sewage sludge. The sludge has to undergo further treatment before being suitable for disposal or application to land.

Sewage treatment may also be referred to as wastewater treatment. However, the latter is a broader term which can also refer to industrial wastewater. For most cities, the sewer system will also carry a proportion of industrial effluent to the sewage treatment plant which has usually received pre-treatment at the factories themselves to reduce the pollutant load. If the sewer system is a combined sewer then it will also carry urban runoff (stormwater) to the sewage treatment plant. Sewage water can travel towards treatment plants via piping and in a flow aided by gravity and pumps. The first part of filtration of sewage typically includes a bar screen to filter solids and large objects which are then collected in dumpsters and disposed of in landfills. Fat and grease is also removed before the primary treatment of sewage.

Sewage is generated by residential, institutional, commercial and industrial establishments. It includes household waste liquid from toilets, baths, showers, kitchens, and sinks draining into sewers. In many areas, sewage also includes liquid waste from industry and commerce. The separation and draining of household waste into greywater and blackwater is becoming more common in the developed world, with treated greywater being permitted to be used for watering plants or recycled for flushing toilets.

Sewage Mixing with Rainwater

Sewage may include stormwater runoff or urban runoff. Sewerage systems capable of handling storm water are known as combined sewer systems. This design was common when urban sewerage systems were first developed, in the late 19th and early 20th centuries. Combined sewers require much larger and more expensive treatment facilities than sanitary sewers. Heavy volumes of storm runoff may overwhelm the sewage treatment system, causing a spill or overflow. Sanitary sewers are typically much smaller than combined sewers, and they are not designed to transport stormwater. Backups of raw sewage can occur if excessive infiltration/ inflow (dilution by stormwater and groundwater) is allowed into a sanitary sewer system. Communities that have urbanized in the mid-20th century or later generally have built separate systems for sewage (sanitary sewers) and stormwater, because precipitation causes widely varying flows, reducing sewage treatment plant efficiency.

As rainfall travels over roofs and the ground, it may pick up various contaminants including soil particles and other sediment, heavy metals, organic compounds, animal waste, and oil and grease. Some jurisdictions require stormwater to receive some level of treatment before being discharged directly into waterways. Examples of treatment processes used for stormwater include retention basins, wetlands, buried vaults with various kinds of media filters, and vortex separators (to remove coarse solids).

Industrial Effluent

In highly regulated developed countries, industrial effluent usually receives at least pretreatment if not full treatment at the factories themselves to reduce the pollutant load, before discharge to the sewer. This process is called industrial wastewater treatment or pretreatment. The same does not apply to many developing countries where industrial effluent is more likely to enter the sewer if it exists, or even the receiving water body, without pretreatment.

Industrial wastewater may contain pollutants which cannot be removed by conventional sewage treatment. Also, variable flow of industrial waste associated with production cycles may upset the population dynamics of biological treatment units, such as the activated sludge process.

Process Steps

Sewage collection and treatment in the United States is typically subject to local, state and federal regulations and standards.

Treating wastewater has the aim to produce an effluent that will do as little harm as possible when discharged to the surrounding environment, thereby preventing pollution compared to releasing untreated wastewater into the environment.

Sewage treatment generally involves three stages, called primary, secondary and tertiary treatment:

- Primary treatment: Consists of temporarily holding the sewage in a quiescent basin where heavy solids can settle to the bottom while oil, grease and lighter solids float to the surface. The settled and floating materials are removed and the remaining liquid may be discharged or subjected to secondary treatment. Some sewage treatment plants that are connected to a combined sewer system have a bypass arrangement after the primary treatment unit. This means that during very heavy rainfall events, the secondary and tertiary treatment systems can be bypassed to protect them from hydraulic overloading, and the mixture of sewage and stormwater only receives primary treatment.
- Secondary treatment: Removes dissolved and suspended biological matter. Secondary treatment is typically performed by indigenous, water-borne micro-organisms in a managed habitat. Secondary treatment may require a separation process to remove the micro-organisms from the treated water prior to discharge or tertiary treatment.

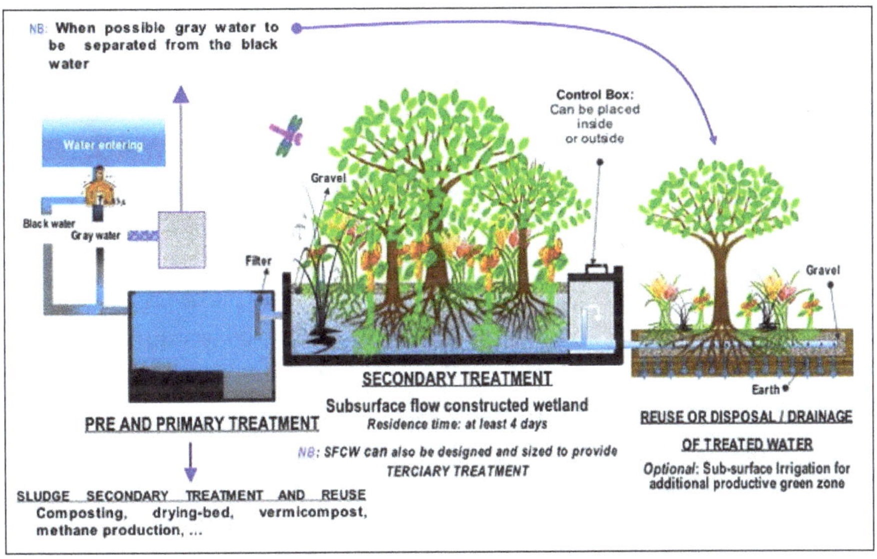

Process flow diagram for a typical treatment plant via subsurface flow constructed wetlands (SFCW).

- Tertiary treatment: Is sometimes defined as anything more than primary and secondary treatment in order to allow ejection into a highly sensitive or fragile ecosystem (estuaries, low-flow rivers, coral reefs). Treated water is sometimes disinfected chemically or physically (for example, by lagoons and microfiltration) prior to discharge into a stream, river, bay, lagoon or wetland, or it can be used for the irrigation of a golf course, green way or park. If it is sufficiently clean, it can also be used for groundwater recharge or agricultural purposes.

Pretreatment

Pretreatment removes all materials that can be easily collected from the raw sewage before they damage or clog the pumps and sewage lines of primary treatment clarifiers. Objects commonly removed during pretreatment include trash, tree limbs, leaves, branches, and other large objects.

The influent in sewage water passes through a bar screen to remove all large objects like cans, rags, sticks, plastic packets etc. carried in the sewage stream. This is most commonly done with an automated mechanically raked bar screen in modern plants serving large populations, while in smaller or less modern plants, a manually cleaned screen may be used. The raking action of a mechanical bar screen is typically paced according to the accumulation on the bar screens and/or flow rate. The solids are collected and later disposed in a landfill, or incinerated. Bar screens or mesh screens of varying sizes may be used to optimize solids removal. If gross solids are not removed, they become entrained in pipes and moving parts of the treatment plant, and can cause substantial damage and inefficiency in the process.

Grit Removal

Grit consists of sand, gravel, cinders, and other heavy materials. It also includes organic matter such as eggshells, bone chips, seeds, and coffee grounds. Pretreatment may include a sand or grit channel or chamber, where the velocity of the incoming sewage is adjusted to allow the settlement of sand and grit. Grit removal is necessary to (1) reduce formation of heavy deposits in aeration tanks, aerobic digesters, pipelines, channels, and conduits; (2) reduce the frequency of digester cleaning caused by excessive accumulations of grit; and (3) protect moving mechanical equipment from abrasion and accompanying abnormal wear. The removal of grit is essential for equipment with closely machined metal surfaces such as comminutors, fine screens, centrifuges, heat exchangers, and high pressure diaphragm pumps. Grit chambers come in 3 types: horizontal grit chambers, aerated grit chambers and vortex grit chambers. Vortex type grit chambers include mechanically induced vortex, hydraulically induced vortex, and multi-tray vortex separators. Given that traditionally, grit removal systems have been designed to remove clean inorganic particles that are greater than 0.210 millimetres (0.0083 in), most grit passes through the grit removal flows under normal conditions. During periods of high flow deposited grit is resuspended and the quantity of grit reaching the treatment plant increases substantially. It is, therefore important that the grit removal system not only operate efficiently during normal flow conditions but also under sustained peak flows when the greatest volume of grit reaches the plant.

Flow Equalization

Clarifiers and mechanized secondary treatment are more efficient under uniform flow conditions. Equalization basins may be used for temporary storage of diurnal or wet-weather flow peaks. Basins provide a place to temporarily hold incoming sewage during plant maintenance and a means of diluting and distributing batch discharges of toxic or high-strength waste which might otherwise inhibit biological secondary treatment (including portable toilet waste, vehicle holding tanks, and septic tank pumpers). Flow equalization basins require variable discharge control, typically include provisions for bypass and cleaning, and may also include aerators. Cleaning may be easier if the basin is downstream of screening and grit removal.

Fat and Grease Removal

In some larger plants, fat and grease are removed by passing the sewage through a small tank where skimmers collect the fat floating on the surface. Air blowers in the base of the tank may also be used to help recover the fat as a froth. Many plants, however, use primary clarifiers with mechanical surface skimmers for fat and grease removal.

Primary Treatment

Primary treatment tanks.

In the primary sedimentation stage, sewage flows through large tanks, commonly called "pre-settling basins", "primary sedimentation tanks" or "primary clarifiers". The tanks are used to settle sludge while grease and oils rise to the surface and are skimmed off. Primary settling tanks are usually equipped with mechanically driven scrapers that continually drive the collected sludge towards a hopper in the base of the tank where it is pumped to sludge treatment facilities. Grease and oil from the floating material can sometimes be recovered for saponification (soap making).

Secondary Treatment

Secondary treatment is designed to substantially degrade the biological content of the sewage which are derived from human waste, food waste, soaps and detergent. The majority of municipal plants treat the settled sewage liquor using aerobic biological processes. To be effective, the biota require both oxygen and food to live. The bacteria and protozoa consume biodegradable soluble organic contaminants (e.g. sugars, fats, organic short-chain carbon molecules, etc.) and bind much of the less soluble fractions into floc. Secondary treatment systems are classified as fixed-film or suspended-growth systems.

- Fixed-film or attached growth systems include trickling filters, constructed wetlands, bio-towers, and rotating biological contactors, where the biomass grows on media and the sewage passes over its surface. The fixed-film principle has further developed into moving bed biofilm reactors (MBBR) and Integrated Fixed-Film Activated Sludge (IFAS) processes. An MBBR system typically requires a smaller footprint than suspended-growth systems.
- Suspended-growth systems include activated sludge, where the biomass is mixed with the sewage and can be operated in a smaller space than trickling filters that treat the same amount of water. However, fixed-film systems are more able to cope with drastic changes in the amount of biological material and can provide higher removal rates for organic material and suspended solids than suspended growth systems.

Secondary clarifier at a rural treatment plant.

Some secondary treatment methods include a secondary clarifier to settle out and separate biological floc or filter material grown in the secondary treatment bioreactor.

Tertiary Treatment

The purpose of tertiary treatment is to provide a final treatment stage to further improve the effluent quality before it is discharged to the receiving environment (sea, river, lake, wet lands, ground, etc.). More than one tertiary treatment process may be used at any treatment plant. If disinfection is practised, it is always the final process. It is also called "effluent polishing".

Filtration

Sand filtration removes much of the residual suspended matter. Filtration over activated carbon, also called *carbon adsorption,* removes residual toxins.

Lagoons or Ponds

A sewage treatment plant and lagoon.

Lagoons or ponds provide settlement and further biological improvement through storage in large man-made ponds or lagoons. These lagoons are highly aerobic and colonization by native macrophytes, especially reeds, is often encouraged. Small filter-feeding invertebrates such as *Daphnia* and species of *Rotifera* greatly assist in treatment by removing fine particulates.

Biological Nutrient Removal

Biological nutrient removal (BNR) is regarded by some as a type of secondary treatment process, and by others as a tertiary (or "advanced") treatment process. Wastewater may contain high levels of the nutrients nitrogen and phosphorus. Excessive release to the environment can lead to a buildup of nutrients, called eutrophication, which can in turn encourage the overgrowth of weeds, algae, and cyanobacteria (blue-green algae). This may cause an algal bloom, a rapid growth in the population of algae. The algae numbers are unsustainable and eventually most of them die. The decomposition of the algae by bacteria uses up so much of the oxygen in the water that most or all of the animals die, which creates more organic matter for the bacteria to decompose. In addition to causing deoxygenation, some algal species produce toxins that contaminate drinking water supplies. Different treatment processes are required to remove nitrogen and phosphorus.

Nitrogen Removal

Nitrogen is removed through the biological oxidation of nitrogen from ammonia to nitrate (nitrification), followed by denitrification, the reduction of nitrate to nitrogen gas. Nitrogen gas is released to the atmosphere and thus removed from the water.

Nitrification itself is a two-step aerobic process, each step facilitated by a different type of bacteria. The oxidation of ammonia (NH_3) to nitrite (NO_2^-) is most often facilitated by *Nitrosomonas* spp. ("nitroso" referring to the formation of a nitroso functional group). Nitrite oxidation to nitrate (NO_3^-), though traditionally believed to be facilitated by *Nitrobacter* spp. (nitro referring the formation of a nitro functional group), is now known to be facilitated in the environment almost exclusively by *Nitrospira* spp.

Denitrification requires anoxic conditions to encourage the appropriate biological communities to form. It is facilitated by a wide diversity of bacteria. Sand filters, lagooning and reed beds can all be used to reduce nitrogen, but the activated sludge process (if designed well) can do the job the most easily. Since denitrification is the reduction of nitrate to dinitrogen (molecular nitrogen) gas, an electron donor is needed. This can be, depending on the waste water, organic matter (from feces), sulfide, or an added donor like methanol. The sludge in the anoxic tanks (denitrification tanks) must be mixed well (mixture of recirculated mixed liquor, return activated sludge [RAS], and raw influent) e.g. by using submersible mixers in order to achieve the desired denitrification.

Sometimes the conversion of toxic ammonia to nitrate alone is referred to as tertiary treatment.

Over time, different treatment configurations have evolved as denitrification has become more sophisticated. An initial scheme, the Ludzack–Ettinger Process, placed an anoxic treatment zone before the aeration tank and clarifier, using the return activated sludge (RAS) from the clarifier as a nitrate source. Influent wastewater (either raw or as effluent from primary clarification) serves as the electron source for the facultative bacteria to metabolize carbon, using the inorganic nitrate as a source of oxygen instead of dissolved molecular oxygen. This denitrification scheme was naturally limited to the amount of soluble nitrate present in the RAS. Nitrate reduction was limited because RAS rate is limited by the performance of the clarifier.

The "Modified Ludzak–Ettinger Process" (MLE) is an improvement on the original concept, for it recycles mixed liquor from the discharge end of the aeration tank to the head of the anoxic tank to provide a consistent source of soluble nitrate for the facultative bacteria. In this instance, raw wastewater continues to provide the electron source, and sub-surface mixing maintains the bacteria in contact with both electron source and soluble nitrate in the absence of dissolved oxygen.

Many sewage treatment plants use centrifugal pumps to transfer the nitrified mixed liquor from the aeration zone to the anoxic zone for denitrification. These pumps are often referred to as *Internal Mixed Liquor Recycle* (IMLR) pumps. IMLR may be 200% to 400% the flow rate of influent wastewater (Q). This is in addition to Return Activated Sludge (RAS) from secondary clarifiers, which may be 100% of Q. (Therefore, the hydraulic capacity of the tanks in such a system should handle at least 400% of annual average design flow (AADF). At times, the raw or primary effluent wastewater must be carbon-supplemented by the addition of methanol, acetate, or simple food waste (molasses, whey, plant starch) to improve the treatment efficiency. These carbon additions should be accounted for in the design of a treatment facility's organic loading. Further modifications to the MLE were to come: Bardenpho and Biodenipho processes include additional anoxic and oxidative processes to further polish the conversion of nitrate ion to molecular nitrogen gas. Use of an anaerobic tank following the initial anoxic process allows for luxury uptake of phosphorus by bacteria, thereby biologically reducing orthophosphate ion in the treated wastewater. Even newer improvements, such as Anammox Process, interrupt the formation of nitrate at the nitrite stage of nitrification, shunting nitrite-rich mixed liquor activated sludge to treatment where nitrite is then converted to molecular nitrogen gas, saving energy, alkalinity, and secondary carbon sourcing. Anammox™ (ANaerobic AMMonia OXidation) works by artificially extending detention time and preserving denitrifiying bacteria through the use of substrate added to the mixed liquor and continuously recycled from it prior to secondary clarification. Many other proprietary schemes are being deployed, including DEMON, Sharon-ANAMMOX™, ANITA-Mox, and DeAmmon. The bacteria Brocadia anammoxidans can remove ammonium from waste water through anaerobic oxidation of ammonium to hydrazine, a form of rocket fuel.

Phosphorus Removal

Every adult human excretes between 200 and 1,000 grams (7.1 and 35.3 oz) of phosphorus annually. Studies of United States sewage in the late 1960s estimated mean per capita contributions of 500 grams (18 oz) in urine and feces, 1,000 grams (35 oz) in synthetic detergents, and lesser variable amounts used as corrosion and scale control chemicals in water supplies. Source control via alternative detergent formulations has subsequently reduced the largest contribution, but the content of urine and feces will remain unchanged. Phosphorus removal is important as it is a limiting nutrient for algae growth in many fresh water systems. It is also particularly important for water reuse systems where high phosphorus concentrations may lead to fouling of downstream equipment such as reverse osmosis.

Phosphorus can be removed biologically in a process called enhanced biological phosphorus removal. In this process, specific bacteria, called polyphosphate-accumulating organisms (PAOs), are selectively enriched and accumulate large quantities of phosphorus within their cells (up to 20 percent of their mass). When the biomass enriched in these bacteria is separated from the treated water, these biosolids have a high fertilizer value.

Phosphorus removal can also be achieved by chemical precipitation, usually with salts of iron (e.g. ferric chloride), aluminum (e.g. alum), or lime. This may lead to excessive sludge production as hydroxides precipitate and the added chemicals can be expensive. Chemical phosphorus removal requires significantly smaller equipment footprint than biological removal, is easier to operate and is often more reliable than biological phosphorus removal. Another method for phosphorus removal is to use granular laterite.

Some systems use both biological phosphorus removal and chemical phosphorus removal. The chemical phosphorus removal in those systems may be used as a backup system, for use when the biological phosphorus removal is not removing enough phosphorus, or may be used continuously. In either case, using both biological and chemical phosphorus removal has the advantage of not increasing sludge production as much as chemical phosphorus removal on its own, with the disadvantage of the increased initial cost associated with installing two different systems.

Once removed, phosphorus, in the form of a phosphate-rich sewage sludge, may be dumped in a landfill or used as fertilizer. In the latter case, the treated sewage sludge is also sometimes referred to as biosolids.

Disinfection

The purpose of disinfection in the treatment of waste water is to substantially reduce the number of microorganisms in the water to be discharged back into the environment for the later use of drinking, bathing, irrigation, etc. The effectiveness of disinfection depends on the quality of the water being treated (e.g., cloudiness, pH, etc.), the type of disinfection being used, the disinfectant dosage (concentration and time), and other environmental variables. Cloudy water will be treated less successfully, since solid matter can shield organisms, especially from ultraviolet light or if contact times are low. Generally, short contact times, low doses and high flows all militate against effective disinfection. Common methods of disinfection include ozone, chlorine, ultraviolet light, or sodium hypochlorite. Chloramine, which is used for drinking water, is not used in the treatment of waste water because of its persistence. After multiple steps of disinfection, the treated water is ready to be released back into the water cycle by means of the nearest body of water or agriculture. Afterwards, the water can be transferred to reserves for everyday human uses.

Chlorination remains the most common form of waste water disinfection in North America due to its low cost and long-term history of effectiveness. One disadvantage is that chlorination of residual organic material can generate chlorinated-organic compounds that may be carcinogenic or harmful to the environment. Residual chlorine or chloramines may also be capable of chlorinating organic material in the natural aquatic environment. Further, because residual chlorine is toxic to aquatic species, the treated effluent must also be chemically dechlorinated, adding to the complexity and cost of treatment.

Ultraviolet (UV) light can be used instead of chlorine, iodine, or other chemicals. Because no chemicals are used, the treated water has no adverse effect on organisms that later consume it, as may be the case with other methods. UV radiation causes damage to the genetic structure of bacteria, viruses, and other pathogens, making them incapable of reproduction. The key disadvantages of UV disinfection are the need for frequent lamp maintenance and replacement and the need for a highly treated effluent to ensure that the target microorganisms are not shielded from the UV radiation (i.e., any solids present in the treated effluent may protect microorganisms from the UV light).

In the United Kingdom, UV light is becoming the most common means of disinfection because of the concerns about the impacts of chlorine in chlorinating residual organics in the wastewater and in chlorinating organics in the receiving water. Some sewage treatment systems in Canada and the US also use UV light for their effluent water disinfection.

Ozone (O_3) is generated by passing oxygen (O_2) through a high voltage potential resulting in a third oxygen atom becoming attached and forming O_3. Ozone is very unstable and reactive and oxidizes most organic material it comes in contact with, thereby destroying many pathogenic microorganisms. Ozone is considered to be safer than chlorine because, unlike chlorine which has to be stored on site (highly poisonous in the event of an accidental release), ozone is generated on-site as needed from the oxygen in the ambient air. Ozonation also produces fewer disinfection by-products than chlorination. A disadvantage of ozone disinfection is the high cost of the ozone generation equipment and the requirements for special operators.

Fourth Treatment Stage

Micropollutants such as pharmaceuticals, ingredients of household chemicals, chemicals used in small businesses or industries, environmental persistent pharmaceutical pollutant (EPPP) or pesticides may not be eliminated in the conventional treatment process (primary, secondary and tertiary treatment) and therefore lead to water pollution. Although concentrations of those substances and their decomposition products are quite low, there is still a chance to harm aquatic organisms. For pharmaceuticals, the following substances have been identified as "toxicologically relevant": substances with endocrine disrupting effects, genotoxic substances and substances that enhance the development of bacterial resistances. They mainly belong to the group of environmental persistent pharmaceutical pollutants. Techniques for elimination of micropollutants via a fourth treatment stage during sewage treatment are implemented in Germany, Switzerland, Sweden and the Netherlands and tests are ongoing in several other countries. Such process steps mainly consist of activated carbon filters that adsorb the micropollutants. The combination of advanced oxidation with ozone followed by GAC, Granulated Activated Carbon, has been suggested as a cost-effective treatment combination for pharmaceutical residues. For a full reduction of microplasts the combination of ultra filtration followed by GAC has been suggested. Also the use of enzymes such as the enzyme laccase is under investigation. A new concept which could provide an energy-efficient treatment of micropollutants could be the use of laccase secreting fungi cultivated at a wastewater treatment plant to degrade micropollutants and at the same time to provide enzymes at a cathode of a microbial biofuel cells. Microbial biofuel cells are investigated for their property to treat organic matter in wastewater.

To reduce pharmaceuticals in water bodies, also "source control" measures are under investigation, such as innovations in drug development or more responsible handling of drugs.

Odor Control

Odors emitted by sewage treatment are typically an indication of an anaerobic or "septic" condition. Early stages of processing will tend to produce foul-smelling gases, with hydrogen sulfide being most common in generating complaints. Large process plants in urban areas will often treat the odors with carbon reactors, a contact media with bio-slimes, small doses of chlorine, or circulating fluids to biologically capture and metabolize the noxious gases. Other methods of odor control exist, including addition of iron salts, hydrogen peroxide, calcium nitrate, etc. to manage hydrogen sulfide levels.

High-density solids pumps are suitable for reducing odors by conveying sludge through hermetic closed pipework.

Energy Requirements

For conventional sewage treatment plants, around 30 percent of the annual operating costs is usually required for energy. The energy requirements vary with type of treatment process as well as wastewater load. For example, constructed wetlands have a lower energy requirement than activated sludge plants, as less energy is required for the aeration step. Sewage treatment plants that produce biogas in their sewage sludge treatment process with anaerobic digestion can produce enough energy to meet most of the energy needs of the sewage treatment plant itself.

In conventional secondary treatment processes, most of the electricity is used for aeration, pumping systems and equipment for the dewatering and drying of sewage sludge. Advanced wastewater treatment plants, e.g. for nutrient removal, require more energy than plants that only achieve primary or secondary treatment.

Sludge Treatment and Disposal

The sludges accumulated in a wastewater treatment process must be treated and disposed of in a safe and effective manner. The purpose of digestion is to reduce the amount of organic matter and the number of disease-causing microorganisms present in the solids. The most common treatment options include anaerobic digestion, aerobic digestion, and composting. Incineration is also used, albeit to a much lesser degree.The use of a green approach, such as phytoremediation, has been recently proposed as a valuable tool to improve sewage sludge contaminated by trace elements and POPs.

Sludge treatment depends on the amount of solids generated and other site-specific conditions. Composting is most often applied to small-scale plants with aerobic digestion for mid-sized operations, and anaerobic digestion for the larger-scale operations.

The sludge is sometimes passed through a so-called pre-thickener which de-waters the sludge. Types of pre-thickeners include centrifugal sludge thickeners rotary drum sludge thickeners and belt filter presses. Dewatered sludge may be incinerated or transported offsite for disposal in a landfill or use as an agricultural soil amendment.

Environment Aspects

Treated water.

Treated water drained to the Elbe river.

Many processes in a wastewater treatment plant are designed to mimic the natural treatment processes that occur in the environment, whether that environment is a natural water body or the ground. If not overloaded, bacteria in the environment will consume organic contaminants, although this will reduce the levels of oxygen in the water and may significantly change the overall ecology of the receiving water. Native bacterial populations feed on the organic contaminants, and the numbers of disease-causing microorganisms are reduced by natural environmental conditions such as predation or exposure to ultraviolet radiation. Consequently, in cases where the receiving environment provides a high level of dilution, a high degree of wastewater treatment may not be required. However, recent evidence has demonstrated that very low levels of specific contaminants in wastewater, including hormones (from animal husbandry and residue from human hormonal contraception methods) and synthetic materials such as phthalates that mimic hormones in their action, can have an unpredictable adverse impact on the natural biota and potentially on humans if the water is re-used for drinking water. In the US and EU, uncontrolled discharges of wastewater to the environment are not permitted under law, and strict water quality requirements are to be met, as clean drinking water is essential. A significant threat in the coming decades will be the increasing uncontrolled discharges of wastewater within rapidly developing countries.

The outlet of the Karlsruhe sewage treatment plant flows into the Alb.

Effects on Biology

Sewage treatment plants can have multiple effects on nutrient levels in the water that the treated sewage flows into. These nutrients can have large effects on the biological life in the water in contact with the effluent. Stabilization ponds (or sewage treatment ponds) can include any of the following:

- Oxidation ponds, which are aerobic bodies of water usually 1–2 metres (3 ft 3 in–6 ft 7 in) in depth that receive effluent from sedimentation tanks or other forms of primary treatment.
    - Dominated by algae.
- Polishing ponds are similar to oxidation ponds but receive effluent from an oxidation pond or from a plant with an extended mechanical treatment.
    - Dominated by zooplankton.
- Facultative lagoons, raw sewage lagoons, or sewage lagoons are ponds where sewage is added with no primary treatment other than coarse screening. These ponds provide effective treatment when the surface remains aerobic; although anaerobic conditions may develop near the layer of settled sludge on the bottom of the pond.
- Anaerobic lagoons are heavily loaded ponds.
    - Dominated by bacteria.
- Sludge lagoons are aerobic ponds, usually 2 to 5 metres (6 ft 7 in to 16 ft 5 in) in depth, that receive anaerobically digested primary sludge, or activated secondary sludge under water.
    - Upper layers are dominated by algae.

Phosphorus limitation is a possible result from sewage treatment and results in flagellate-dominated plankton, particularly in summer and fall.

A phytoplankton study found high nutrient concentrations linked to sewage effluents. High nutrient concentration leads to high chlorophyll a concentrations, which is a proxy for primary production in marine environments. High primary production means high phytoplankton populations and most likely high zooplankton populations, because zooplankton feed on phytoplankton. However, effluent released into marine systems also leads to greater population instability.

The planktonic trends of high populations close to input of treated sewage is contrasted by the bacterial trend. In a study of *Aeromonas* spp. in increasing distance from a wastewater source, greater change in seasonal cycles was found the furthest from the effluent. This trend is so strong that the furthest location studied actually had an inversion of the *Aeromonas* spp. cycle in comparison to that of fecal coliforms. Since there is a main pattern in the cycles that occurred simultaneously at all stations it indicates seasonal factors (temperature, solar radiation, phytoplankton) control of the bacterial population. The effluent dominant species changes from *Aeromonas caviae* in winter to *Aeromonas sobria* in the spring and fall while the inflow dominant species is *Aeromonas caviae*, which is constant throughout the seasons.

Reuse

With suitable technology, it is possible to reuse sewage effluent for drinking water, although this is usually only done in places with limited water supplies, such as Windhoek and Singapore.

In arid countries, treated wastewater is often used in agriculture. For example, in Israel, about 50 percent of agricultural water use (total use was one billion cubic metres (3.5×10^{10} cu ft)) is provided through reclaimed sewer water. Future plans call for increased use of treated sewer water as well as more desalination plants as part of water supply and sanitation in Israel.

Constructed wetlands fed by wastewater provide both treatment and habitats for flora and fauna. Another example for reuse combined with treatment of sewage are the East Kolkata Wetlands in India. These wetlands are used to treat Kolkata's sewage, and the nutrients contained in the wastewater sustain fish farms and agriculture.

Developing Countries

Few reliable figures exist on the share of the wastewater collected in sewers that is being treated in the world. A global estimate by UNDP and UN-Habitat is that 90% of all wastewater generated is released into the environment untreated. In many developing countries the bulk of domestic and industrial wastewater is discharged without any treatment or after primary treatment only.

Sewage treatment plant in Gobabis, Namibia.

In Latin America about 15 percent of collected wastewater passes through treatment plants (with varying levels of actual treatment). In Venezuela, a below average country in South America with respect to wastewater treatment, 97 percent of the country's sewage is discharged raw into the environment.

In Iran, a relatively developed Middle Eastern country, the majority of Tehran's population has totally untreated sewage injected to the city's groundwater. However, the construction of major parts of the sewage system, collection and treatment, in Tehran is almost complete, and under development, due to be fully completed by the end of 2012. In Isfahan, Iran's third largest city, sewage treatment was started more than 100 years ago.

Only few cities in sub-Saharan Africa have sewer-based sanitation systems, let alone wastewater treatment plants, an exception being South Africa and – until the late 1990s – Zimbabwe. Instead, most urban residents in sub-Saharan Africa rely on on-site sanitation systems without sewers, such as septic tanks and pit latrines, and fecal sludge management in these cities is an enormous challenge.

Industrial Wastewater Treatment

Industrial wastewater treatment describes the processes used for treating wastewater that is produced by industries as an undesirable by-product. After treatment, the treated industrial wastewater (or effluent) may be reused or released to a sanitary sewer or to a surface water in the environment.

Most industries produce some wastewater. Recent trends have been to minimize such production or to recycle treated wastewater within the production process.

Sources of Industrial Wastewater

Battery Manufacturing

Battery manufacturers specialize in fabricating small devices for electronics and portable equipment (e.g., power tools), or larger, high-powered units for cars, trucks and other motorized vehicles. Pollutants generated at manufacturing plants includes cadmium, chromium, cobalt, copper, cyanide, iron, lead, manganese, mercury, nickel, oil & grease, silver and zinc.

Organic Chemicals Manufacturing

The specific pollutants discharged by organic chemical manufacturers vary widely from plant to plant, depending on the types of products manufactured, such as bulk organic chemicals, resins, pesticides, plastics, or synthetic fibers. Some of the organic compounds that may be discharged are benzene, chloroform, napthalene, phenols, toluene and vinyl chloride. Biochemical oxygen demand (BOD), which is a gross measurement of a range of organic pollutants, may be used to gauge the effectiveness of a biological wastewater treatment system, and is used as a regulatory parameter in some discharge permits. Metal pollutant discharges may include chromium, copper, lead, nickel and zinc.

Electric Power Plants

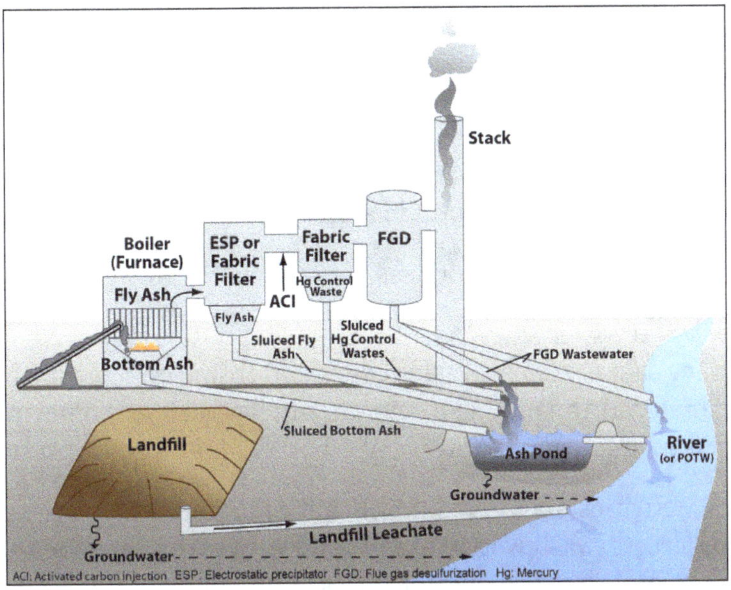

Wastestreams at a coal-fired power plant.

Fossil-fuel power stations, particularly coal-fired plants, are a major source of industrial wastewater. Many of these plants discharge wastewater with significant levels of metals such as lead, mercury, cadmium and chromium, as well as arsenic, selenium and nitrogen compounds (nitrates and nitrites). Wastewater streams include flue-gas desulfurization, fly ash, bottom ash and flue gas mercury control. Plants with air pollution controls such as wet scrubbers typically transfer the captured pollutants to the wastewater stream.

Ash ponds, a type of surface impoundment, are a widely used treatment technology at coal-fired plants. These ponds use gravity to settle out large particulates (measured as total suspended solids) from power plant wastewater. This technology does not treat dissolved pollutants. Power stations use additional technologies to control pollutants, depending on the particular wastestream in the plant. These include dry ash handling, closed-loop ash recycling, chemical precipitation, biological treatment (such as an activated sludge process), membrane systems, and evaporation-crystallization systems. Technological advancements in ion exchange membranes and electrodialysis systems has enabled high efficiency treatment of flue-gas desulfurization wastewater to meet recent EPA discharge limits. The treatment approach is similar for other highly scaling industrial wastewaters.

Food Industry

Wastewater generated from agricultural and food processing operations has distinctive characteristics that set it apart from common municipal wastewater managed by public or private sewage treatment plants throughout the world: it is biodegradable and non-toxic, but has high concentrations of BOD and suspended solids (SS). The constituents of food and agriculture wastewater are often complex to predict, due to the differences in BOD and pH in effluents from vegetable, fruit, and meat products and due to the seasonal nature of food processing and post-harvesting.

Processing of food from raw materials requires large volumes of high grade water. Vegetable washing generates waters with high loads of particulate matter and some dissolved organic matter. It may also contain surfactants.

Animal slaughter and processing produces very strong organic waste from body fluids, such as blood, and gut contents.

Processing food for sale produces wastes generated from cooking which are often rich in plant organic material and may also contain salt, flavourings, colouring material and acids or alkali. Very significant quantities of oil or fats may also be present.

Food processing activities such as plant cleaning, material conveying, bottling, and product washing create wastewater. Many food processing facilities require on-site treatment before operational wastewater can be land applied or discharged to a waterway or a sewer system. High suspended solids levels of organic particles increase BOD and can result in significant sewer surcharge fees. Sedimentation, wedgewire screening, or rotating belt filtration (microscreening) are commonly used methods to reduce suspended organic solids loading prior to discharge.

Iron and Steel Industry

The production of iron from its ores involves powerful reduction reactions in blast furnaces. Cooling waters are inevitably contaminated with products especially ammonia and cyanide. Production of coke from coal in coking plants also requires water cooling and the use of water in by-products separation. Contamination of waste streams includes gasification products such as benzene, naphthalene, anthracene, cyanide, ammonia, phenols, cresols together with a range of more complex organic compounds known collectively as polycyclic aromatic hydrocarbons (PAH).

The conversion of iron or steel into sheet, wire or rods requires hot and cold mechanical transformation stages frequently employing water as a lubricant and coolant. Contaminants include hydraulic oils, tallow and particulate solids. Final treatment of iron and steel products before onward sale into manufacturing includes *pickling* in strong mineral acid to remove rust and prepare the surface for tin or chromium plating or for other surface treatments such as galvanisation or painting. The two acids commonly used are hydrochloric acid and sulfuric acid. Wastewaters include acidic rinse waters together with waste acid. Although many plants operate acid recovery plants (particularly those using hydrochloric acid), where the mineral acid is boiled away from the iron salts, there remains a large volume of highly acid ferrous sulfate or ferrous chloride to be disposed of. Many steel industry wastewaters are contaminated by hydraulic oil, also known as *soluble oil.*

Mines and Quarries

Mine wastewater effluent in Peru, with neutralized pH from tailing runoff.

The principal waste-waters associated with mines and quarries are slurries of rock particles in water. These arise from rainfall washing exposed surfaces and haul roads and also from rock washing and grading processes. Volumes of water can be very high, especially rainfall related arisings on large sites. Some specialized separation operations, such as coal washing to separate coal from native rock using density gradients, can produce wastewater contaminated by fine particulate haematite and surfactants. Oils and hydraulic oils are also common contaminants.

Wastewater from metal mines and ore recovery plants are inevitably contaminated by the minerals present in the native rock formations. Following crushing and extraction of the desirable materials, undesirable materials may enter the wastewater stream. For metal mines, this can include unwanted metals such as zinc and other materials such as arsenic. Extraction of high value metals such as gold and silver may generate slimes containing very fine particles in where physical removal of contaminants becomes particularly difficult.

Additionally, the geologic formations that harbour economically valuable metals such as copper and gold very often consist of sulphide-type ores. The processing entails grinding the rock into fine particles and then extracting the desired metal(s), with the leftover rock being known as tailings. These tailings contain a combination of not only undesirable leftover metals, but also sulphide components which eventually form sulphuric acid upon the exposure to air and water that inevitably occurs when the tailings are disposed of in large impoundments. The resulting acid mine drainage, which is often rich in heavy metals (because acids dissolve metals), is one of the many environmental impacts of mining.

Nuclear Industry

The waste production from the nuclear and radio-chemicals industry is dealt with as *Radioactive waste.*

Petroleum Refining and Petrochemicals

Pollutants discharged at petroleum refineries and petrochemical plants include conventional pollutants (BOD, oil and grease, suspended solids), ammonia, chromium, phenols and sulfides.

Pulp and Paper Industry

Effluent from the pulp and paper industry is generally high in suspended solids and BOD. Plants that bleach wood pulp for paper making may generate chloroform, dioxins (including 2,3,7,8-TCDD), furans, phenols and chemical oxygen demand (COD). Stand-alone paper mills using imported pulp may only require simple primary treatment, such as sedimentation or dissolved air flotation. Increased BOD or COD loadings, as well as organic pollutants, may require biological treatment such as activated sludge or upflow anaerobic sludge blanket reactors. For mills with high inorganic loadings like salt, tertiary treatments may be required, either general membrane treatments like ultrafiltration or reverse osmosis or treatments to remove specific contaminants, such as nutrients.

Textile Dyeing

Textile dyeing plants generate wastewater that contain synthetic and natural dyestuff, gum thickener (guar) and various wetting agents, pH buffers and dye retardants or accelerators. Following treatment with polymer-based flocculants and settling agents, typical monitoring parameters include BOD, COD, color (ADMI), sulfide, oil and grease, phenol, TSS and heavy metals (chromium, zinc, lead, copper).

Industrial Oil Contamination

Industrial applications where oil enters the wastewater stream may include vehicle wash bays, workshops, fuel storage depots, transport hubs and power generation. Often the wastewater is discharged into local sewer or trade waste systems and must meet local environmental specifications. Typical contaminants can include solvents, detergents, grit. lubricants and hydrocarbons.

Water Treatment

Many industries have a need to treat water to obtain very high quality water for demanding purposes such pure chemical synthesis or boiler feed water. Many water treatment produce organic and mineral sludges from filtration and sedimentation. Ion exchange using natural or synthetic resins removes calcium, magnesium and carbonate ions from water, typically replacing them with sodium, chloride, hydroxyl and/or other ions. Regeneration of ion exchange columns with strong acids and alkalis produces a wastewater rich in hardness ions which are readily precipitated out, especially when in admixture with other wastewater constituents.

Wood Preserving

Wood preserving plants generate conventional and toxic pollutants, including arsenic, COD, copper, chromium, abnormally high or low pH, phenols, oil & grease, and suspended solids.

Wool Processing

Insecticide residues in fleeces are a particular problem in treating waters generated in wool processing. Animal fats may be present in the wastewater, which if not contaminated, can be recovered for the production of tallow or further rendering.

Treatment of Industrial Wastewater

The various types of contamination of wastewater require a variety of strategies to remove the contamination.

Brine Treatment

Brine treatment involves removing dissolved salt ions from the waste stream. Although similarities to seawater or brackish water desalination exist, industrial brine treatment may contain unique combinations of dissolved ions, such as hardness ions or other metals, necessitating specific processes and equipment.

Brine treatment systems are typically optimized to either reduce the volume of the final discharge for more economic disposal (as disposal costs are often based on volume) or maximize the recovery of fresh water or salts. Brine treatment systems may also be optimized to reduce electricity consumption, chemical usage, or physical footprint.

Brine treatment is commonly encountered when treating cooling tower blowdown, produced water from steam assisted gravity drainage (SAGD), produced water from natural gas extraction such as coal seam gas, frac flowback water, acid mine or acid rock drainage, reverse osmosis reject, chlor-alkali wastewater, pulp and paper mill effluent, and waste streams from food and beverage processing.

Brine treatment technologies may include: membrane filtration processes, such as reverse osmosis; ion exchange processes such as electrodialysis or weak acid cation exchange; or evaporation processes, such as brine concentrators and crystallizers employing mechanical vapour recompression and steam.

Reverse osmosis may not be viable for brine treatment, due to the potential for fouling caused by hardness salts or organic contaminants, or damage to the reverse osmosis membranes from hydrocarbons.

Evaporation processes are the most widespread for brine treatment as they enable the highest degree of concentration, as high as solid salt. They also produce the highest purity effluent, even distillate-quality. Evaporation processes are also more tolerant of organics, hydrocarbons, or hardness salts. However, energy consumption is high and corrosion may be an issue as the prime mover is concentrated salt water. As a result, evaporation systems typically employ titanium or duplex stainless steel materials.

Brine Management

Brine management examines the broader context of brine treatment and may include consideration of government policy and regulations, corporate sustainability, environmental impact, recycling, handling and transport, containment, centralized compared to on-site treatment, avoidance and reduction, technologies, and economics. Brine management shares some issues with leachate management and more general waste management.

Solids Removal

Most solids can be removed using simple sedimentation techniques with the solids recovered as slurry or sludge. Very fine solids and solids with densities close to the density of water pose special problems. In such case filtration or ultrafiltration may be required. Although, flocculation may be used, using alum salts or the addition of polyelectrolytes. Wastewater from industrial food processing often requires on-site treatment before it can be discharged to prevent or reduce sewer surcharge fees. The type of industry and specific operational practices determine what types of wastewater is generated and what type of treatment is required. Reducing solids such as waste product, organic materials, and sand is often a goal of industrial wastewater treatment. Some common ways to reduce solids include primary sedimentation (clarification), Dissolved Air Flotation or (DAF), belt filtration (microscreening), and drum screening.

Oils and Grease Removal

The effective removal of oils and grease is dependent on the characteristics of the oil in terms of its suspension state and droplet size, which will in turn affect the choice of separator technology. Oil in industrial waste water may be free light oil, heavy oil, which tends to sink, and emulsified oil, often referred to as soluble oil. Emulsified or soluble oils will typically required "cracking" to free the oil from its emulsion. In most cases this is achieved by lowering the pH of the water matrix.

Most separator technologies will have an optimum range of oil droplet sizes that can be effectively treated.

Analysing the oily water to determine droplet size can be performed with a video particle analyser. Each separator technology will have its own performance curve outlining optimum performance based on oil droplet size. the most common separators are gravity tanks or pits, API oil-water separators or plate packs, chemical treatment via DAFs, centrifuges, media filters and hydrocyclones.

API Separators

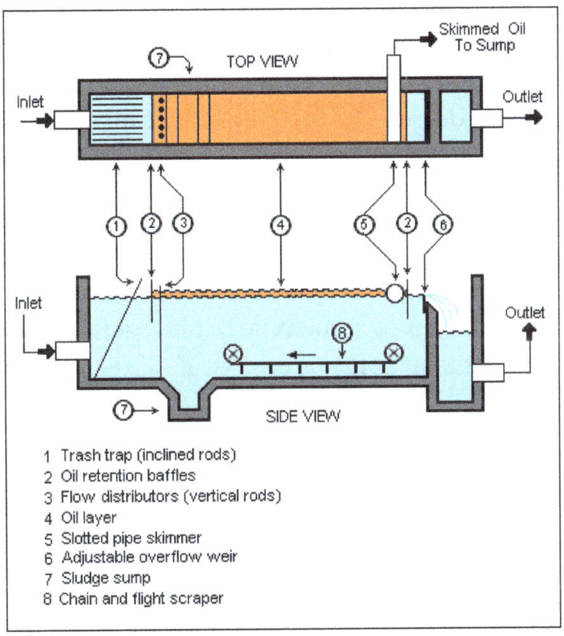

A typical API oil-water separator used in many industries.

Many oils can be recovered from open water surfaces by skimming devices. Considered a dependable and cheap way to remove oil, grease and other hydrocarbons from water, oil skimmers can sometimes achieve the desired level of water purity. At other times, skimming is also a cost-efficient method to remove most of the oil before using membrane filters and chemical processes. Skimmers will prevent filters from blinding prematurely and keep chemical costs down because there is less oil to process.

Because grease skimming involves higher viscosity hydrocarbons, skimmers must be equipped with heaters powerful enough to keep grease fluid for discharge. If floating grease forms into solid clumps or mats, a spray bar, aerator or mechanical apparatus can be used to facilitate removal.

However, hydraulic oils and the majority of oils that have degraded to any extent will also have a soluble or emulsified component that will require further treatment to eliminate. Dissolving or emulsifying oil using surfactants or solvents usually exacerbates the problem rather than solving it, producing wastewater that is more difficult to treat.

The wastewaters from large-scale industries such as oil refineries, petrochemical plants, chemical plants, and natural gas processing plants commonly contain gross amounts of oil and suspended solids. Those industries use a device known as an API oil-water separator which is designed to separate the oil and suspended solids from their wastewater effluents. The name is derived from the fact that such separators are designed according to standards published by the American Petroleum Institute (API).

The API separator is a gravity separation device designed by using Stokes Law to define the rise velocity of oil droplets based on their density and size. The design is based on the specific gravity difference between the oil and the wastewater because that difference is much smaller than the specific gravity difference between the suspended solids and water. The suspended solids settles to the bottom of the separator as a sediment layer, the oil rises to top of the separator and the cleansed wastewater is the middle layer between the oil layer and the solids.

Typically, the oil layer is skimmed off and subsequently re-processed or disposed of, and the bottom sediment layer is removed by a chain and flight scraper (or similar device) and a sludge pump. The water layer is sent to further treatment for additional removal of any residual oil and then to some type of biological treatment unit for removal of undesirable dissolved chemical compounds.

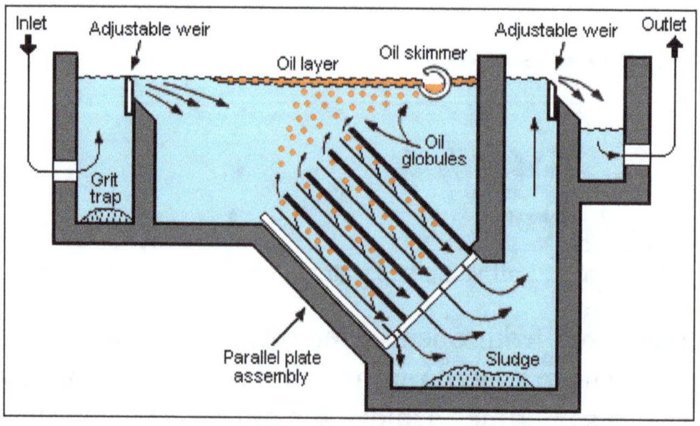

A typical parallel plate separator.

Parallel plate separators are similar to API separators but they include tilted parallel plate assemblies (also known as parallel packs). The parallel plates provide more surface for suspended oil droplets to coalesce into larger globules. Such separators still depend upon the specific gravity between the suspended oil and the water. However, the parallel plates enhance the degree of oil-water separation. The result is that a parallel plate separator requires significantly less space than a conventional API separator to achieve the same degree of separation.

Hydrocyclone Oil Separators

Hydrocyclone oil separators operate on the process where wastewater enters the cyclone chamber and is spun under extreme centrifugal forces more than 1000 times the force of gravity. This force causes the water and oil droplets to separate. The separated oil is discharged from one end of the cyclone where treated water is discharged through the opposite end for further treatment, filtration or discharge.

Removal of Biodegradable Organics

Biodegradable organic material of plant or animal origin is usually possible to treat using extended conventional sewage treatment processes such as activated sludge or trickling filter. Problems can arise if the wastewater is excessively diluted with washing water or is highly concentrated such as undiluted blood or milk. The presence of cleaning agents, disinfectants, pesticides, or antibiotics can have detrimental impacts on treatment processes.

Activated Sludge Process

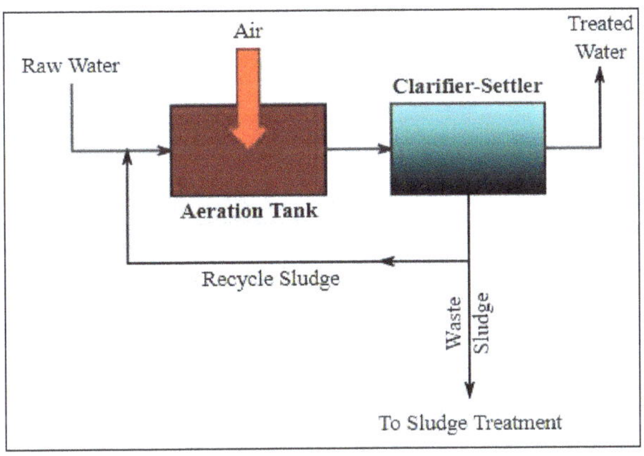

A generalized diagram of an activated sludge process.

Activated sludge is a biochemical process for treating sewage and industrial wastewater that uses air (or oxygen) and microorganisms to biologically oxidize organic pollutants, producing a waste sludge (or floc) containing the oxidized material. In general, an activated sludge process includes:

- An aeration tank where air (or oxygen) is injected and thoroughly mixed into the wastewater.

- A settling tank (usually referred to as a clarifier or "settler") to allow the waste sludge to settle. Part of the waste sludge is recycled to the aeration tank and the remaining waste sludge is removed for further treatment and ultimate disposal.

Trickling Filter Process

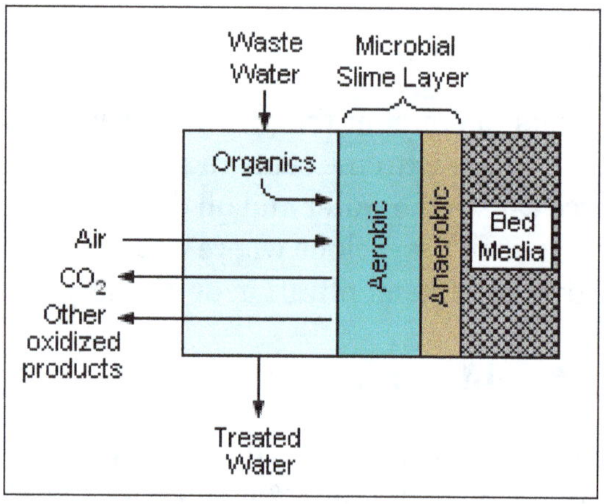

Image: A schematic cross-section of the contact face of the bed media in a trickling filter.

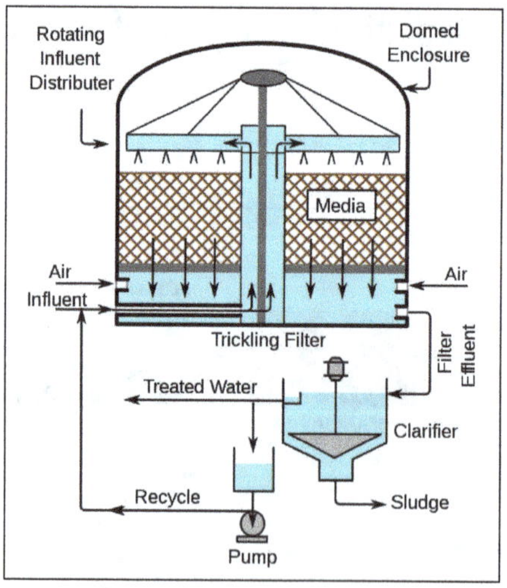

A typical complete trickling filter system.

A trickling filter consists of a bed of rocks, gravel, slag, peat moss, or plastic media over which wastewater flows downward and contacts a layer (or film) of microbial slime covering the bed media. Aerobic conditions are maintained by forced air flowing through the bed or by natural convection of air. The process involves adsorption of organic compounds in the wastewater by the microbial slime layer, diffusion of air into the slime layer to provide the oxygen required for the biochemical oxidation of the organic compounds. The end products include carbon dioxide gas, water and other products of the oxidation. As the slime layer thickens, it becomes difficult for the air to penetrate the layer and an inner anaerobic layer is formed.

The fundamental components of a complete trickling filter system are:

- A bed of filter medium upon which a layer of microbial slime is promoted and developed.
- An enclosure or a container which houses the bed of filter medium.
- A system for distributing the flow of wastewater over the filter medium.
- A system for removing and disposing of any sludge from the treated effluent.

The treatment of sewage or other wastewater with trickling filters is among the oldest and most well characterized treatment technologies.

A trickling filter is also often called a trickle filter, trickling biofilter, biofilter, biological filter or biological trickling filter.

Removal of other Organics

Synthetic organic materials including solvents, paints, pharmaceuticals, pesticides, products from coke production and so forth can be very difficult to treat. Treatment methods are often specific to the material being treated. Methods include advanced oxidation processing, distillation, adsorption, ozonation, vitrification, incineration, chemical immobilisation or landfill disposal. Some materials such as some detergents may be capable of biological degradation and in such cases, a modified form of wastewater treatment can be used.

Removal of Acids and Alkalis

Acids and alkalis can usually be neutralised under controlled conditions. Neutralisation frequently produces a precipitate that will require treatment as a solid residue that may also be toxic. In some cases, gases may be evolved requiring treatment for the gas stream. Some other forms of treatment are usually required following neutralisation.

Waste streams rich in hardness ions as from de-ionisation processes can readily lose the hardness ions in a buildup of precipitated calcium and magnesium salts. This precipitation process can cause severe *furring* of pipes and can, in extreme cases, cause the blockage of disposal pipes. A 1-metre diameter industrial marine discharge pipe serving a major chemicals complex was blocked by such salts in the 1970s. Treatment is by concentration of de-ionisation waste waters and disposal to landfill or by careful pH management of the released wastewater.

Removal of Toxic Materials

Toxic materials including many organic materials, metals (such as zinc, silver, cadmium, thallium, etc.) acids, alkalis, non-metallic elements (such as arsenic or selenium) are generally resistant to biological processes unless very dilute. Metals can often be precipitated out by changing the pH or by treatment with other chemicals. Many, however, are resistant to treatment or mitigation and may require concentration followed by landfilling or recycling. Dissolved organics can be *incinerated* within the wastewater by the advanced oxidation process.

Smart Capsules

Molecular encapsulation is a technology that has the potential to provide a system for the recyclable removal of lead and other ions from polluted sources. Nano-, micro- and milli- capsules, with sizes in the ranges 10 nm-1µm,1µm-1mm and >1mm, respectively, are particles that have an active reagent (core) surrounded by a carrier (shell).There are three types of capsule under investigation: alginate-based capsules, carbon nanotubes, polymer swelling capsules. These capsules provide a possible means for the remediation of contaminated water.

Noise Pollution Control

Noise pollution can be effectively controlled by taking the following measures:

Control at Receiver's End

For people working in noisy installations, ear-protection aids like ear-plugs, ear-muffs, noise helmets, headphones etc. must be provided to reduce occupational exposure.

Suppression of Noise at Source

This is possible if working methods are improved by:

- Designing, fabricating and using quieter machines to replace the noisy ones.
- Proper lubrication and better maintenance of machines.
- Installing noisy machines in sound proof chambers.
- Covering noise-producing machine parts with sound-absorbing materials to check noise production.
- Reducing the noise produced from a vibrating machine by vibration damping i.e. making a layer of damping material (rubber, neoprene, cork or plastic) beneath the machine.
- Using silencers to control noise from automobiles, ducts, exhausts etc. and convey systems with ends opening into the atmosphere.

- Using glass wool or mineral wool covered with a sheet of perforated metal for the purpose of mechanical protection.

Acoustic Zoning

Increased distance between source and receiver by zoning of noisy industrial areas, bus terminals and railway stations, aerodromes etc. away from the residential areas would go a long way in minimising noise pollution. There should be silence zones near the residential areas, educational institutions and above all, near hospitals.

Sound Insulation at Construction Stages

- Sound travels through the cracks that get left between the door and the wall. For reducing noise, this space (jamb frame gap) should be packed with sound absorbing material.
- Sound insulation can be done by constructing windows with double or triple panes of glass and filling the gaps with sound absorbing materials.
- Acoustical tiles, hair felt, perforated plywood etc. can be fixed on walls, ceilings, floors etc. to reduce noise (especially for sound proof recording rooms etc.).

Planting of Trees

Planting green trees and shrubs along roads, hospitals, educational institutions etc. help in noise reduction to a considerable extent.

Legislative Measures

Strict legislative measures need to be enforced to curb the menace of noise pollution. Some of these measures could be:

- Minimum use of loudspeakers and amplifiers especially near silence zones.
- Banning pressure horns in automobiles.
- Framing a separate Noise Pollution Act.

Reference

- Pollution-control, technology: britannica.com, Retrieved 5 February, 2020
- Khopkar, S.M. (2004). Environmental Pollution Monitoring And Control. New Delhi: New Age International. P. 299. ISBN 978-81-224-1507-0
- Air-pollution-control, technology: britannica.com, Retrieved 30 April, 2020
- Tchobanoglous, George; Burton, Franklin L.; Stensel, H. David; Metcalf & Eddy, Inc. (2003). Wastewater Engineering: Treatment and Reuse (4th ed.). Mcgraw-Hill. ISBN 978-0-07-112250-4

- "Sanitation Systems – Sanitation Technologies – Activated sludge". SSWM. 27 April 2018. Retrieved 31 October 2018
- Tchobanoglous, G., Burton, F.L., and Stensel, H.D. (2003). Wastewater Engineering (Treatment Disposal Reuse) / Metcalf & Eddy, Inc (4th ed.). Mcgraw-Hill Book Company. ISBN 0-07-041878-0
- How-to-control-noise-pollution-6-effective-measures, noise-pollution: yourarticlelibrary.com, Retrieved 22 May, 2020

PERMISSIONS

All chapters in this book are published with permission under the Creative Commons Attribution Share Alike License or equivalent. Every chapter published in this book has been scrutinized by our experts. Their significance has been extensively debated. The topics covered herein carry significant information for a comprehensive understanding. They may even be implemented as practical applications or may be referred to as a beginning point for further studies.

We would like to thank the editorial team for lending their expertise to make the book truly unique. They have played a crucial role in the development of this book. Without their invaluable contributions this book wouldn't have been possible. They have made vital efforts to compile up to date information on the varied aspects of this subject to make this book a valuable addition to the collection of many professionals and students.

This book was conceptualized with the vision of imparting up-to-date and integrated information in this field. To ensure the same, a matchless editorial board was set up. Every individual on the board went through rigorous rounds of assessment to prove their worth. After which they invested a large part of their time researching and compiling the most relevant data for our readers.

The editorial board has been involved in producing this book since its inception. They have spent rigorous hours researching and exploring the diverse topics which have resulted in the successful publishing of this book. They have passed on their knowledge of decades through this book. To expedite this challenging task, the publisher supported the team at every step. A small team of assistant editors was also appointed to further simplify the editing procedure and attain best results for the readers.

Apart from the editorial board, the designing team has also invested a significant amount of their time in understanding the subject and creating the most relevant covers. They scrutinized every image to scout for the most suitable representation of the subject and create an appropriate cover for the book.

The publishing team has been an ardent support to the editorial, designing and production team. Their endless efforts to recruit the best for this project, has resulted in the accomplishment of this book. They are a veteran in the field of academics and their pool of knowledge is as vast as their experience in printing. Their expertise and guidance has proved useful at every step. Their uncompromising quality standards have made this book an exceptional effort. Their encouragement from time to time has been an inspiration for everyone.

The publisher and the editorial board hope that this book will prove to be a valuable piece of knowledge for students, practitioners and scholars across the globe.

INDEX